OBSTETRICS AND GYNECOLOGY ADVANCES

HYALURONIC ACID

ROLE IN PREGNANCY AND NOVEL APPLICATIONS IN THE GESTATIONAL PERIOD

OBSTETRICS AND GYNECOLOGY ADVANCES

Additional books and e-books in this series can be found on Nova's website under the Series tab.

OBSTETRICS AND GYNECOLOGY ADVANCES

HYALURONIC ACID

ROLE IN PREGNANCY AND NOVEL APPLICATIONS IN THE GESTATIONAL PERIOD

VITTORIO UNFER
EDITOR

Copyright © 2021 by Nova Science Publishers, Inc.

All rights reserved. No part of this book may be reproduced, stored in a retrieval system or transmitted in any form or by any means: electronic, electrostatic, magnetic, tape, mechanical photocopying, recording or otherwise without the written permission of the Publisher.

We have partnered with Copyright Clearance Center to make it easy for you to obtain permissions to reuse content from this publication. Simply navigate to this publication's page on Nova's website and locate the "Get Permission" button below the title description. This button is linked directly to the title's permission page on copyright.com. Alternatively, you can visit copyright.com and search by title, ISBN, or ISSN.

For further questions about using the service on copyright.com, please contact:
Copyright Clearance Center
Phone: +1-(978) 750-8400   Fax: +1-(978) 750-4470   E-mail: info@copyright.com.

NOTICE TO THE READER

The Publisher has taken reasonable care in the preparation of this book, but makes no expressed or implied warranty of any kind and assumes no responsibility for any errors or omissions. No liability is assumed for incidental or consequential damages in connection with or arising out of information contained in this book. The Publisher shall not be liable for any special, consequential, or exemplary damages resulting, in whole or in part, from the readers' use of, or reliance upon, this material. Any parts of this book based on government reports are so indicated and copyright is claimed for those parts to the extent applicable to compilations of such works.

Independent verification should be sought for any data, advice or recommendations contained in this book. In addition, no responsibility is assumed by the Publisher for any injury and/or damage to persons or property arising from any methods, products, instructions, ideas or otherwise contained in this publication.

This publication is designed to provide accurate and authoritative information with regard to the subject matter covered herein. It is sold with the clear understanding that the Publisher is not engaged in rendering legal or any other professional services. If legal or any other expert assistance is required, the services of a competent person should be sought. FROM A DECLARATION OF PARTICIPANTS JOINTLY ADOPTED BY A COMMITTEE OF THE AMERICAN BAR ASSOCIATION AND A COMMITTEE OF PUBLISHERS.

Additional color graphics may be available in the e-book version of this book.

Library of Congress Cataloging-in-Publication Data

ISBN: 978-1-53619-743-3

Published by Nova Science Publishers, Inc. † New York

Hyaluronic acid has multiple functions in health and disease. Due to its wide range of functions and properties, hyaluronic acid is a versatile molecule widely used in different sectors. Physiologically, HA is one of the chief components of the extracellular matrix, expressed in the main organs and tissues involved in pregnancy in all stages of gestation, from fertilization to implantation, from fetal development to finally parturition. This book gives an interesting overview about the pivotal role played by High Molecular Hyaluronic Acid (HMW-HA) during pregnancy. Scientific evidence on the benefic effects displayed by HMW-HA in tissues that participate to reproductive function are promptly collected to help professionals and researchers who wants to acquire more information on several issues related to the supportive care of pregnancy by the aid of natural molecules. Moreover, the high safety profile of HMW-HA supports its safety administration in pregnancy, representing a valid contribution in gestational complications.

Prof. Imelda Hernàndez Marin M.D.
Specialist in Human Reproduction
Universidad Nacional Autónoma de México (UNAM)

This volume provides a valuable new insight into novel aspects of Hyaluronic Acid (HA), mostly by focusing on the role that HA plays during pregnancy and other gynecological conditions.
Hyaluronic acid is an anionic, non-sulfated glycosaminoglycan distributed widely throughout connective, epithelial, and neural tissues, which exerts relevant tasks also in modulating the endometrium, namely during pregnancy.

As one of the chief components of the extracellular matrix, HA contributes to driving cell migration and in shaping the architecture of the cell microenvironment. This aspect has been carefully investigated during the last decade, given that changes in HA concentration and/or structure may confer significant properties to the microenvironment, by impairing/fostering cell plasticity, invasiveness, and differentiation. Moreover, as highlighted by studies performed on human skin, HA is involved in repair processes and in those tissue changes that occur downstream of inflammation. HA modulate a number of cell surface receptor interactions, notably those including HA, CD44 and RHAMM receptors. It is worth of interest that the interaction between HA and CD44 may modulate cell adhesion interactions and, by this way, can affect cell motility and invasiveness in both pathological (cancer) and physiological (oocyte transfer, embryo attachment, pregnancy) conditions. Furthermore, HA degradation products can amplify the inflammatory response by transducing the inflammatory signal through toll-like receptors in macrophage and other immunocompetent cells. The book collects a number of chapters dealing with the effects displayed by HA in tissues that participate to reproductive function in which HA has demonstrated to favor the physiological gestation. It is likely that such an effect could be ascribed to the microenvironment-related modifications. Indeed, HA enhances blastocyst implantation while modulating the reactive immunological response triggered by embryo adhesion. Overall, those effects, in some way, mimic a progesterone-like activity. Those evidence prompted to suggest a supportive role for HA in the management of complicated pregnancy as well as in the adjuvant treatment of several gynecological-related conditions.

The book – written by experts in the field – may significantly help both students and physicians – in acquiring a more in dept awareness of several issues related to the supportive care of pregnancy by the aid of natural molecules.

Prof. Mariano Bizzarri, M.D., PhD
Professor of Pathology in the Department of Experimental Medicine
University La Sapienza
Roma, Italy

The investigation of Hyaluronan (HA) involvement in pregnancy and throughout the gestational period has been an increasing area of investigation and speculation for over 30 years. With the growing body of information and our greater understanding, admittedly still incomplete, comes the opportunity for application and perhaps designing interventions with this very special, molecularly simple, multifunctional molecule.

HA, despite being a biochemically simple linear carbohydrate molecule consisting of disaccharide repeats without chemical modifications, and having no protein component, has multiple profound functions in health and disease in vertebrate animals, including humans. Nowhere are HA's myriad functions, and regulation of its metabolism, more obviously important than during the period that encompasses fertilization, implantation, fetal development and finally parturition. Even upon birth, mother's milk provides HA that likely strengthens innate host defense mechanisms in the newborn. Many of HA's effects are orchestrated by binding to one or more of a number specific cell receptors, or by forming a complex extracellular matrix incorporating other matricellular proteoglycans/proteins. Through these mechanisms information is contributed to the surrounding cell populations, and appears to be critical for the timing of embryo development and delivery. Therefore the frank discussion of the current state of the art, dissection of unanswered questions, as well as imagining how the information could lead to application when the biological processes are compromised, is a quite timely. Commercial biosynthesis of highly pure HA, produced to medical use standards, has already been achieved and can facilitate safe translation. HA's lack of immunogenicity, and normally wide distribution within vertebrates makes it a very attractive treatment modality once efficacy is demonstrated.

Carol de la Motte, PhD
Department of Inflammation and Immunity
Lerner Research Institute, Cleveland Clinic Foundation
Cleveland, OH, United States

CONTENTS

Preface		xi
Chapter 1	Properties and Physiological Role of Hyaluronic Acid *Marco Tilotta, Gianpiero Forte and Sara Proietti*	1
Chapter 2	Hyaluronic Acid in Obstetrics: Role in Physiological Pregnancy *Fabio Facchinetti and Vittorio Unfer*	23
Chapter 3	Effect of Hyaluronic Acid Treatment in Maintaining the Physiological Pregnancy: Preclinical Evidence *Serap Cilaker Micili and Asli Goker*	53
Chapter 4	Safety of Hyaluronic Acid in Pregnancy *Giovanni Buzzaccarini, Marco Noventa and Antonio Simone Laganà*	65
Chapter 5	Use of Hyaluronic Acid in Assisted Reproduction Techniques *Maria Salomé Bezerra Espinola, Berniero Visconti and Cesare Aragona*	83
Chapter 6	Role of Hyaluronan in Fetal Development *Cora M. Demler and Natasza A. Kurpios*	105

Chapter 7	Hyaluronic Acid in the Development of the Gut and Protection against Necrotizing Enterocolitis *Kathryn Y. Burge, Jeffrey V. Eckert* *and Hala Chaaban*	**143**

About the Editors   **171**

Index   **175**

PREFACE

This book offers a valuable clinical resource for health professionals and researchers. It gives an overview about the pivotal role played by High Molecular Hyaluronic Acid (HMW-HA) during pregnancy and its applications in the gestational period. Hyaluronic acid is a critical component of the extracellular matrix (ECM) and one of the most interesting, versatile and useful natural molecules in almost all areas of biology. As widely reported, HA has a pivotal role in several phases of pregnancy, from fertilization to labour and it displays several regulatory activities and functional properties based on its different molecular weight. While low molecular weight hyaluronic acid (LMW-HA) is widely used in gynaecology for menopause-related symptoms or in association with physical treatments (e.g., radiation therapy) to counteract the onset of adverse events, HMW-HA has been poorly evaluated as clinical treatment. Its physiological presence in the extracellular matrix of all the main organs and tissues involved in pregnancy (uterus, cervix, placenta, decidua, chorion, amnios, ovarium etc..) suggests the importance of this molecule for a successful gestation. Several papers, in particular, shed light on its importance for blastocyst adhesion and implantation, for an efficient immune tolerance and for the correct development of haemo-lymphatic system. It has been demonstrated that HMW-HA has regulatory activity on the PGRMC1 expression, a specific progesterone receptor expressed in maternal and foetal-maternal interface tissues, involved in uterine quiescence. On these premises, the administration of HMW-HA may represent an interesting treatment opportunity for the prevention of recurrent miscarriage and pre-term birth (PTB) in patients with risk factors. The high safety profile of HMW-HA further supports its administration in pregnancy. All these interesting topics will be discussed and deepened in this book, giving the opportunity to explore in detail every aspect of this effective and attractive molecule, thus helping physicians to assess the state of connective tissues in pregnancy and to evaluate the risk for the onset of gestational complications.

In: Hyaluronic Acid
Editor: Vittorio Unfer
ISBN: 978-1-53619-743-3
© 2021 Nova Science Publishers, Inc.

Chapter 1

PROPERTIES AND PHYSIOLOGICAL ROLE OF HYALURONIC ACID

Marco Tilotta, Gianpiero Forte and Sara Proietti[*]
R&D Department, Lo.Li. Pharma srl, Rome, Italy

ABSTRACT

Hyaluronic Acid (hyaluronan, HA) is a linear polysaccharide formed from disaccharide units containing N-acetyl-D-glucosamine and glucuronic acid. It is a critical component of the extracellular matrix (ECM) and one of the most interesting, versatile and useful natural molecules in almost all areas of biology. The progressive understanding of HA's biological roles and its peculiar features made it possible to use hyaluronic acid in different fields, ranging from medical to pharmaceutical, nutritional and cosmetic application. Though its simple primary structure, HA regulates biological responses in a highly complex manner principally depending on its concentration and molecular weight which seems to be the key for its pleiotropic functions. HA size influences its

---

[*] Corresponding Author's Email: s.proietti@lolipharma.it.

affinity to receptors thus affecting its uptake by the cell and modulating different biological responses.

Keywords: hyaluronan, hyaluronan synthase, hyaluronidases, high molecular weight hyaluronan, low molecular weight hyaluronan

INTRODUCTION

Hyaluronic Acid (HA) is one of the most interesting, versatile and useful natural macromolecules that, though its simple chemical structure, has extraordinary properties and plays important roles in almost all areas of biology. It is the principal glycosaminoglycan (GAG) in the skin (Juhlin 1997), the mucous membranes and body fluids such as synovial fluid (Hamerman and Schuster 1958) vitreous humor, and Wharton's jelly of the umbilical cord. Due to its chemo-physical properties, it is widely involved at the structural and regulatory levels (Ialenti and Di Rosa 1994). Physiologically, HA occurs as a salt form (hyaluronate) of high molecular weight hyaluronic acid (HMW-HA) with high concentrations in several connective soft tissues. Significant amounts of HA are also found in lung, kidney, brain, and muscle tissues (Fraser, Laurent, and Laurent 1997).

Hyaluronic Acid was purified for the first time in 1934 when Karl Meyer and John Palmer isolated it from the vitreous humor of the bovine eye and named it HA, "hyalos" from the Greek word for glass and uronic acid (Meyer 1934). During the 1930s and 1950s, HA was isolated also from the human umbilical cord, rooster comb and streptococci (Danishefsky and Bella 1966, Dawson 1937). Since then, a variety of studies evidenced how HA is broadly diffused in nature, present in all tissues and body fluids of vertebrates as well as in some bacteria.

Its structure was widely studied from the 1940s (Kaye and Stacey 1950, Blumberg et al. 1958) and it was solved in 1954 by Meyer and Weissmann (Weissmann 1954).

In 1986 Endre Balazs introduced for the first time the term "hyaluronan" to conform it with the international nomenclature of polysaccharides (Balazs, Laurent, and Jeanloz 1986).

The progressive understanding of HA's biological roles and peculiar features made it possible to use hyaluronic acid in different fields, ranging from medical to pharmaceutical, nutritional and cosmetic application. HA has a beneficial treatment of joint (Bergstrand S 2019, Cooper et al. 2017), skin disease (Pavicic et al. 2011) and soft tissue augmentation (Moon et al. 2019, Prasetyo et al. 2016).

Maintenance of the elastoviscosity of liquid connective tissues, control of tissue hydration, and supramolecular assembly of proteoglycans represent the main activities of this molecule. Moreover, evidence suggest that HA also plays important roles in the organization of extracellular matrix (ECM) and immunity response, through specific and non-specific interactions (Fraser, Laurent, and Laurent 1997). HA is a well-known fundamental component for skin integrity, where it acts as a barrier that incorporates water molecules and reduces the transition of other small molecules, excluding macromolecules from the extracellular matrix by steric effect (Ogston and Phelps 1961). Due to its high viscosity, HA can contribute to hamper the passage of viruses and bacteria through the pericellular zone (Clarris, Fraser, and Rodda 1974, Clarris and Fraser 1968). In addition, HA plays a pivotal role in the wound healing process by stimulating fibroblast proliferation, remodeling of ECM and keratinocyte migration (Chen and Abatangelo 1999, Sahana and Rekha 2018, Neuman et al. 2015).

Currently, HA is used in several other branches of medicine (pulmonology, orthopedics, aesthetic medicine, gynecology, ophthalmology, etc.) without contraindications or reported interactions with drugs (Jentsch et al. 2003, Nolan et al. 2006, Rodriguez-Merchan 2013, Migliore et al. 2017, A Ciofalo 2017). The versatility of this molecule is related to its molecular weight. HA can have different molecular weights and each of them allows HA interaction with several receptors thus regulating different biological processes or different phases of the same process (Garantziotis and Savani 2019).

HYALURONIC ACID STRUCTURE

HA belongs to the family of glycosaminoglycans (GAGs), which includes heparan sulphate and chondroitin sulphate. It is a negatively charged, high molecular weight polysaccharide composed of repeating of N-acetyl-D-glucosamine (GlcNac) and D-glucuronic acid (GlcA) bound through alternating ß-1,4 and ß-1,3 glycosidic bonds (Atkins and Sheehan 1971). As other GAGs, HA is one component of the extracellular matrix (ECM), a complex network of macromolecules existing within all tissues and organs that provides essential physical scaffolding for the cellular constituents and initiates crucial biochemical and biomechanical cues required for tissue morphogenesis, differentiation and homeostasis (Frantz, Stewart, and Weaver 2010).

Though it is classified as a GAG, HA differs from them due to its highly variable molecular weight: the number of repeating disaccharides can largely vary so that the molecule mass ranges from 2 to 10,000 or more kDa (from 10^4 to 10^7 Da) (Itano et al. 1999). Due to this characteristic, HA can be defined as oligo-HA (O-HA: molecular weight <10 kDa), low molecular weight HA (LMW-HA: molecular weight ranging from 10 to 500 kDa), and high molecular weight HA (HMW-HA >500 kDa). The concentration and size distribution of HA differs with tissue type, age and disease severity (Cowman et al. 2015). Unlike other GAGs, HA is not synthesized by resident Golgi enzymes and covalently attached to core proteins, but it is synthesized at the inner face of the plasma membrane as a free linear polymer (Toole 2004).

HYALURONAN SYNTHESIS

HA has multiple physical and physiological properties depending on its weight and concentration and a fine balance between synthesis and degradation allows to maintain HA homeostasis and a high turnover rate (Tammi, Day, and Turley 2002) (Figure 1).

	O-HA (oligomeric HA)	LMW-HA (low molecular weight HA)	HMW-HA (high molecular weight HA)
	Molecular mass 0.75-10kDa (2-20 disaccharide units)	Molecular mass <500kDa (~1000 disaccharide units or less)	Molecular mass >500kDa (~2000-25000 disaccharide units)
FIELD OF APPLICATION	URO/GYNECOLOGY	GYNECOLOGY	OBSTETRICS
INDICATION	Bladder dysfunction, vaginal dryness, vaginal atrophy, urinary incontinence, cystitis	Treatment and prevention of vaginal diseases (lesions, ectopias...) and other alterations of the vaginal mucosa caused by physical treatments (DTC, laser therapy, cryotherapy, radiotherapy). Soothing and lubricant treatment, especially in menopausal patients	Adjuvant in conditions in which it is necessary to maintain normal physiological processes in obstetric-gynecological clinical practice
ACTION	Regulation of various processes both positively and negatively	Anti-inflammatory action, pro-angiogenic effect, stimulation of endothelial cell proliferation, adhesion and formation of capillaries, synthesis of pro-inflammatory cytokines, stimulation of apoptosis of decidual stromal cells in early pregnancy, stimulation of extravillous trophoblast cell invasion	Formation and stabilization of ECM, immunologically inert, anti-inflammatory action, anti-angiogenic effect, synthesis of immunosuppressive cytokines, stimulation of proliferation and inhibition of apoptosis of decidual stromal cells in early pregnancy
ABSORPTION	INTESTINAL: permeation through the intestinal barrier passing passively between the enterocytes in the circulatory system. CAECUM	VAGINAL MUCOSA	INTESTINAL: through microfold cells (M cells) enters lymphatic system; through enterocytes transported to blood circulation. COLON

Figure 1. Biological roles of HA based on different molecular weights and primary fields of application

In humans, HA biosynthesis is tightly regulated by the activity of three transmembrane glycosyltransferase isoenzymes hyaluronan synthases: HAS1, HAS2 and HAS3. The three isoforms have a central hydrophilic region and a catalytic site on the inner face of the plasma membrane, that utilizes uridine diphosphate glucuronic acid (UDP-GlcUA) and uridine diphosphate N-acetylglucosamine (UDP-GlcNAc) as substrates to form HA polymers. Biochemical analyses revealed that each HAS isoform differs in terms of activity, product elongation rate and stability (Itano et al. 1999). Although HAS gene sequences are located on different chromosome (hCh19-HAS1, hCh8-HAS2 and hCh16-HAS3) (Spicer and McDonald 1998), the three HAS isoforms expression and activity are controlled in a specific manner with different regulatory systems (Vigetti et al. 2014). Of particular interest is the finding that these three HAS enzymes synthesize HA of varying molecular mass. HAS1 produces HMW hyaluronan (from 2×10^5 to 2×10^6 Da), while HAS2 is more active and synthesizes HA chains greater than 2×10^6 Da. It represents the main hyaluronan synthetic enzyme in normal adult cells, and its activity is finely regulated (Vigetti et al. 2014). HAS3 produces HA molecules with MW lower than 3×10^5 Da (Girish and Kemparaju 2007).

The HAS genes exhibit different temporal patterns of expression during morphogenesis (Dicker et al. 2014). HAS2 is expressed throughout all stages of embryogenesis (Tien and Spicer 2005) and is considered to be the major hyaluronan synthase during development. Even if the exact regulation mechanisms of each HAS isoenzyme have not been elucidated yet, alterations in HAS gene expression result in an abnormal production of HA, lead to an increased risk in pathological events, alter cell responses to injury and aberrant biological process (Heldin et al. 2019), thus demonstrating how HA synthases are critical mediators in development, injury, and disease.

HYALURONAN CATABOLISM

HA catabolism depends on the activity of specific endoglycosidases called hyaluronidases (HYAL) and a non-specific mechanism due to reactive oxygen species (ROS) including superoxide, hydrogen peroxide, nitric oxide and peroxynitrite, and hypohalous acids (Soltes et al. 2006). Together, these two different mechanisms degrade 30% of the total content of HA in the human body. The remaining 70% is transported by the lymph to the lymph nodes, where it is internalized and catabolized by the endothelial cells of the lymphatic vessels. Additionally, a small part of HA is carried to the bloodstream and degraded by liver endothelial cells (Heldin et al. 2019). In mammals HYAL family includes six enzymes: HYAL1, HYAL2, and HYAL3 genes clustered on human chromosome 3p21.3; HYAL4, HYALP1 and PH-20/sperm adhesion molecule 1 (SPAM1) genes located on chromosome 7p31.3 (Kobayashi, Chanmee, and Itano 2020). HYALs are highly homologous endoglycosidases that specifically hydrolyze the β-1,4 linkage of the HA molecule. Of the 6 HYAL family members, Hyal1 and Hyal2 hyaluronidases are the predominant isoforms that catabolize HA in somatic tissues mainly at acidic pH (Csoka, Frost, and Stern 2001); Hyal3 is weakly expressed in a wide range of somatic cells and does not play a key role in constitutive HA degradation

(Atmuri et al. 2008) with low levels in brain, liver, testis, and bone marrow (Triggs-Raine et al. 1999); Hyal4, which was recently identified as a chondroitin sulphate (CS)-specific hydrolase, has no activity on HA (Kaneiwa et al. 2010), with expression in the placenta and skeletal muscle (Csoka, Frost, and Stern 2001); PH-20/SPAM1, a glycosylphosphatidylinositol (GPI)-anchored hyaluronidase, plays an essential role in fertilization (Martin-Deleon 2011). HMW hyaluronan can also be naturally degraded in the organism by ROS, including superoxide, hydrogen peroxide, nitric oxide, peroxynitrite and hypohalous acids, which are massively produced during inflammatory responses, tissue injury and tumorigenesis (Soltes et al. 2006).

HYALURONAN RECEPTORS

Hyaluronan interacts with cell surfaces in two different ways. First, it can bind to specific receptors to induce a wide range of intracellular signals either directly or by activating other receptors (Turley, Noble, and Bourguignon 2002). Second, HA can be retained at the cell surface generating a voluminous "coat" that incorporates other hyaluronan binding molecules (Toole 2001).

Alterations in ECM structure due to pathological conditions (tumor invasion, inflammation, tissue remodeling, etc.) can induce fragmentation of endogenous HMW-HA by hyaluronidases (Girish and Kemparaju 2007) and reactive oxygen species (Soltes et al. 2006) into LMW-HA, then depolymerized to O-HA. HA and its degradation products bind to several cell surface receptors (Hardwick et al. 1992, Prevo et al. 2001).

CD44 is the best characterized HA receptor. It is present in almost all human cells, detected in various segments of the reproductive system both in humans (Campbell et al. 1995) and in other species (Bergqvist et al. 2005, Perry K 2010, Furnus CC 2003) under physiological conditions. It is a single-pass transmembrane glycoprotein with four functional domains (Ponta, Sherman, and Herrlich 2003) expressed in different isoforms. Alternate splicing, variations in its polypeptide sequence, glycosylation or oligomerization influence its affinity for HA (Ponta, Wainwright, and Herrlich 1998). The interaction between HMW-HA and CD44 controls the expression of several genes that regulate cell growth/survival, cytoskeletal rearrangements, membrane ruffling, and lead to active cell migration (Misra et al. 2015, Tzircotis, Thorne, and Isacke 2005) through extracellular regulated kinase (ERK), phosphoinositide 3-kinase (PI3K), Rac and Ras in various cell types (Kothapalli et al. 2008).

RHAMM receptor (Receptor for Hyaluronan Mediated Motility), also known as CD168, exits in different spliced forms on the cell surface, within the cytoplasm, and in the nucleus. RHAMM is expressed by several cell types and it is involved in different cell signalling controlling cytoskeletal organization (Assmann V 1999).

LYVE-1 (Lymphatic Vessel Endothelial Hyaluronan Receptor 1) is also an important HA receptor. LYVE 1 is a type I integral membrane glycoprotein with a strong homology with CD44. Despite the typical expression in the lymphatic system, (Wróbel T 2005), where its main activity is the transport of HA from the tissues to the lymph by uptake via lymphatic endothelial cells (Prevo et al. 2001), this receptor was also detected in sinusoidal endothelial cells in the liver (Mouta Carreira C 2001).

TLR2 and TLR4 (Toll-like receptors) are two other important receptors binding to HA. They are involved in the recognition of bacterial lipopolysaccharides and lipopeptides during the immune response. TLR4 is a transmembrane receptor able to recognize LPS molecules, whereas TLR2 can recognize mycobacteria and gram positive bacteria (Takeuchi et al. 1999).

TLR2 and TLR4 are normally expressed on the membrane of dendritic cells, monocytes and lymphocytes, and HA interacts with them depending on their molecular weight. Specifically, LMW-HA binds to TLR2 and TLR4, inducing an inflammatory reaction necessary for wound healing process (Jiang, Liang, and Noble 2011, Cyphert, Trempus, and Garantziotis 2015).

HARE (Hyaluronan Receptor for Endocytosis) is a 190 kDa polypeptide generated by the proteolytic cleavage of stabilin 2 or FEEL2 (Harris et al. 2007). HARE protein is described in the endothelial cells in the liver, spleen and lymph nodes, and also from different tissues, including the eye, brain, kidney and heart. HARE receptors mediate hyaluronan endocytosis and are co-localized on cell membrane of endothelial cells with clathrin. HARE is not only specific for hyaluronan, but it is also able to recognize other GAGs (Zhou et al. 2002) (Figure 2).

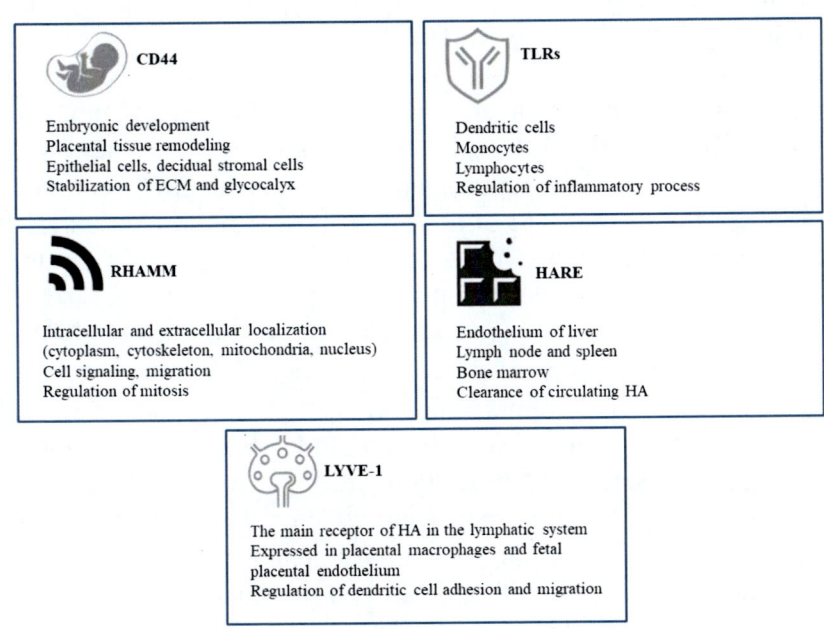

Figure 2. HA receptors and their principal function.

MOLECULAR WEIGHT AND DIFFERENT BIOLOGICAL ROLES

Hyaluronan has a key role in several biological processes in different cell types and tissues, both in normal and pathological conditions. It is well known that this polymer can modulate many biological effects and its regulatory function not only depends on the equilibrium between HA synthesis and degradation, but it is mainly due to its size. Indeed, high molecular weight (HMW) and low molecular weight (LMW) hyaluronan can have even opposite effects (Cyphert, Trempus, and Garantziotis 2015). HMW-HA is a fundamental component of skin integrity. Due to its extremely hydrophilic nature, it acts as a barrier incorporating water molecules and reducing the transition of other small molecules (Ogston AG 1961, Ogston and Sherman 1961). Moreover, due to its high viscosity, HA can contribute to hinder the passage of viruses and bacteria through the pericellular zone (Clarris, Fraser, and Rodda 1974).

Endogenous HMW-HA has anti-inflammatory and antiangiogenic properties, and it is able to inhibit endothelial cell growth (Jiang, Liang, and Noble 2011). HMW-HA can maintain epithelial cell integrity and promote recovery from acute lung injury (Jiang et al. 2005) in a sepsis model of ventilated rats (Liu et al. 2008) as well as in ozone-induced airway hyperreactivity (AHR) (Garantziotis et al. 2009). Due to its viscoelasticity, it acts as a lubricant agent in the synovial joint fluid, thus protecting the articular cartilage (Tamer 2013). Moreover, HMW-HA has also important and beneficial roles in repair, wound healing (Voigt and Driver 2012) and immunosuppression: it binds fibrinogen and controls the recruitment of inflammatory cells, the levels of inflammatory cytokines and the migration of stem cells (Jiang, Liang, and Noble 2011). During some environmental and pathological conditions, such as asthma, pulmonary fibrosis and hypertension, chronic obstructive pulmonary disease and rheumatoid arthritis, HMW-HA is cleaved into LMW-HA (2×10^4–10^6 Da), which acts as endogenous danger signals with pro-inflammatory and pro-angiogenic activities. Indeed, LMW hyaluronan is able to stimulate the production of proinflammatory cytokines, chemokines and growth factors (Cyphert, Trempus, and Garantziotis 2015) and to promote ECM remodeling (Heldin et al. 2019).

ABSORPTION, DISTRIBUTION AND EXCRETION OF HIGH MOLECULAR WEIGHT HYALURONIC ACID

Over the last years, oral administration of exogenous HA has gained attention as a supplementary therapy to prevent or treat local and systemic diseases. The molecular weight, beside being critical in the various biological functions exerted by hyaluronic acid, also determines its fate once ingested.

HA permeation through the intestinal barrier depends on different factors, as cell microenvironment, HA concentration and molecular weight.

Experimental studies evidence that O-HA can permeate through the intestinal barrier passively by passing between the enterocytes in the circulatory system (Kimura et al. 2016).

On the contrary, HA with a molecular weight >10kDA is initially exposed to the acidic environment of the gastric system where, depending on its molecular weight, undergoes pH-induced degradation. As evidenced by Maleki et al. (Maleki, Kjøniksen, and Nyström 2008) the degradation effect is faster at pH 13 than at pH 4 and it is MW dependent. In fact, unlike LMW-HA, high molecular weight hyaluronic acid (HMW-HA >10^5 kDa) does not undergo appreciable degradation during its transit in the gastric system (Kimura et al. 2016, Turley 2003). While HA with lower molecular weight can be degraded into oligosaccharides and then absorbed in the caecum, HA >300 kDa is not degraded by artificial gastric or artificial intestinal juices (Kimura et al. 2016).

Thickness and pore size of mucus nanostructure prevent the free transit of molecules and microorganisms from luminal to the basolateral site, but there is an activate internalization of macromolecules (Linden et al. 2008). Few studies have investigated about transepithelial uptake of HMM-HA ≥ 10^5 Da, and the conclusions are sometimes controversial. Some authors evaluated HA absorption with radiolabeled molecules. Among them, Balogh et al. (Balogh et al. 2008) investigated HA (1000 kDa) absorption using 99Technetium (99Tc) labeling in orally administered rats and dogs. Results showed a rapid absorption and excretion of HA in urine, with accumulation in the thyroid gland, stomach, kidney and, bladder. The absorption of undegraded HMW-HA can occur by epithelial cells as enterocytes (Neal et al. 2006), microfold (M) cells (Barthe, Woodley, and Houin 1999), dendritic cells and macrophages (Corr, Gahan, and Hill 2008), and the TLR4 receptor mediates such process (Oe et al. 2016). In this mechanism of absorption, only M cells deliver intact HMW-HA to the gut-associated lymphatic tissue (Rubas and Grass 1991). Indeed, once internalized by enterocytes or immune cells, HMW-HA is partially digested in O-HA or LMW-HA, and transported to blood circulation within lysosomes. On the contrary, when absorbed by M cells, HMW-HA enters the lymphatic system unaltered and reaches all body districts (Balogh et al. 2008) (Figure 3).

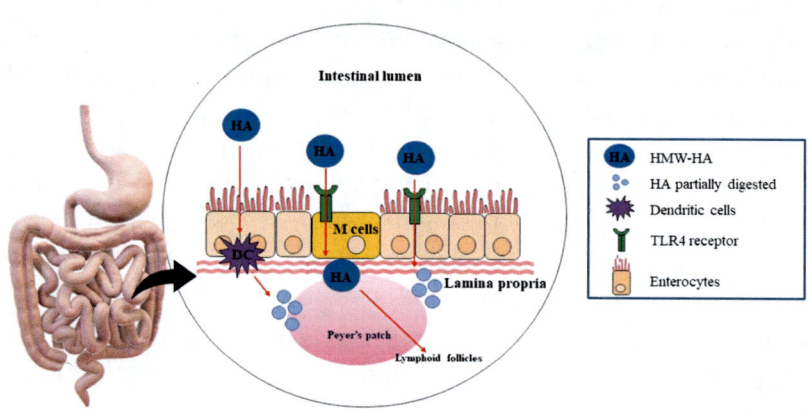

Figure 3. High molecular weight hyaluronic acid (HMW-HA) absorption ways.

CONCLUSION

Hyaluronic acid has a pivotal role in several biological processes. Its pleiotropic functions and versatility are related to its molecular weight and subsequent interactions with a wide range of receptors, leading to activation of different pathways. HMW-HA is widely involved both in structural and regulatory activities, including maintenance of the elastoviscosity, control of tissue hydration, supramolecular assembly of proteoglycans, modulation of the immune response. Moreover, HMW-HA acts as an anti-inflammatory molecule. LMW-HA, on the contrary, is mainly involved in the wound healing process, stimulating immunity response and increasing angiogenesis. The recent acquisition of data related to the intestinal absorption of HMW-HA supports its potential therapeutical application as a treatment for several conditions connected with impaired HMW-HA metabolism, shedding light on the new potential therapeutical application of oral use of this molecule. Future investigations and technological improvements are required, but clinical treatments with HMW-HA may be considered a major and promising innovation.

REFERENCES

Assmann, V., Jenkinson, D., Marshall, J. F. & Hart, I. R. (1999). "The intracellular receptor RHAMM/IHABP interacts with microtubules and actin filaments." *J Cell Sci*, 112, 3943-54.

Atkins, E. D. & Sheehan, J. K. (1971). "The molecular structure of hyaluronic acid." *Biochem J*, 125 (4), 92P. doi: 10.1042/bj1250092pb.

Atmuri, V., Martin, D. C., Hemming, R., Gutsol, A., Byers, S., Sahebjam, S., Thliveris, J. A., Mort, J. S., Carmona, E., Anderson, J. E., Dakshinamurti, S. & Triggs-Raine, B. (2008). "Hyaluronidase 3 (HYAL3) knockout mice do not display evidence of hyaluronan accumulation." *Matrix Biol*, 27 (8), 653-60. doi: 10.1016/j.matbio.2008.07.006.

Balazs, E. A., Laurent, T. C. & Jeanloz, R. W. (1986). "Nomenclature of hyaluronic acid." *Biochem J*, 235 (3), 903. doi: 10.1042/bj2350903.

Balogh, L., Polyak, A., Mathe, D., Kiraly, R., Thuroczy, J., Terez, M., Janoki, G., Ting, Y., Bucci, L. R. & Schauss, A. G. (2008). "Absorption, uptake and tissue affinity of high-molecular-weight hyaluronan after oral administration in rats and dogs." *J Agric Food Chem*, 56 (22), 10582-93. doi: 10.1021/jf8017029.

Barthe, L., Woodley, J. & Houin, G. (1999). "Gastrointestinal absorption of drugs: methods and studies." *Fundam Clin Pharmacol*, 13 (2), 154-68. doi: 10.1111/j.1472-8206.1999.tb00334.x.

Bergqvist, A. S., Yokoo, M., Bage, R., Sato, E. & Rodriguez-Martinez, H. (2005). "Detection of the hyaluronan receptor CD44 in the bovine oviductal epithelium." *J Reprod Dev*, 51 (4), 445-53. doi: 10.1262/jrd.17010.

Bergstrand, S., Ingstad, H. K., Møystad, A. & Bjørnland, T. J. (2019). "Long-term effectiveness of arthrocentesis with and without hyaluronic acid injection for treatment of temporomandibular joint osteoarthritis." *J Oral Sci.* doi: 10.2334/josnusd.

Blumberg, B. S., Ogston, A. G., Lowther, D. A. & Rogers, H. J. (1958). "Physicochemical properties of hyaluronic acid formed by Streptococcus haemolyticus." *Biochem J*, 70 (1), 1-4. doi: 10.1042/bj0700001.

Campbell, S., Swann, H. R., Aplin, J. D., Seif, M. W., Kimber, S. J. & Elstein, M. (1995). "CD44 is expressed throughout pre-implantation human embryo development." *Hum Reprod*, 10 (2), 425-30. doi: 10.1093/oxfordjournals.humrep.a135955.

Chen, W. Y. & Abatangelo, G. (1999). "Functions of hyaluronan in wound repair." *Wound Repair Regen*, 7 (2), 79-89. doi: 10.1046/j.1524-475x.1999.00079.x.

Ciofalo, A., Zambetti, G., Altissimi, G., Fusconi, M., Soldo, P., Gelardi, M., Iannella, G., Pasquariello, B. & Magliulo, G. (2017). "Pathological and cytological changes of the nasal mucosa in acute rhinosinusitis: the role of hyaluronic acid as supportive therapy." *Eur Rev Med Pharmacol Sci.*

Clarris, B. J. & Fraser, J. R. E. (1968). "On the pericellular zone of some mammalian cells *in vitro*." *Experimental Cell Research*, 49 (1), 181-193. doi: 10.1016/0014-4827(68)90530-2.

Clarris, B. J., Fraser, J. R. & Rodda, S. (1974). "Effect of cell-bound hyaluronic acid on infectivity of Newcastle disease virus for human synovial cells *in vitro*." *Ann Rheum Dis*, 33 (3), 240-2. doi: 10.1136/ard.33.3.240.

Cooper, C., Rannou, F., Richette, P., Bruyere, O., Al-Daghri, N., Altman, R. D., Brandi, M. L., Collaud Basset, S., Herrero-Beaumont, G., Migliore, A., Pavelka, K., Uebelhart, D. & Reginster, J. Y. (2017). "Use of Intraarticular Hyaluronic Acid in the Management of Knee Osteoarthritis in Clinical Practice." *Arthritis Care Res (Hoboken)*, 69 (9), 1287-1296. doi: 10.1002/acr.23204.

Corr, S. C., Gahan, C. C. & Hill, C. (2008). "M-cells: origin, morphology and role in mucosal immunity and microbial pathogenesis." *FEMS*

Immunol Med Microbiol, 52 (1), 2-12. doi: 10.1111/j.1574-695X. 2007.00359.x.

Cowman, M. K., Lee, H. G., Schwertfeger, K. L., McCarthy, J. B. & Turley, E. A. (2015). "The Content and Size of Hyaluronan in Biological Fluids and Tissues." *Front Immunol*, 6, 261. doi: 10.3389/fimmu.2015.00261.

Csoka, Antonei B., Gregory I. Frost. & Robert Stern. (2001). "The six hyaluronidase-like genes in the human and mouse genomes." *Matrix Biology*, 20 (8), 499-508. doi: 10.1016/s0945-053x(01)00172-x.

Cyphert, J. M., Trempus, C. S. & Garantziotis, S. (2015). "Size Matters: Molecular Weight Specificity of Hyaluronan Effects in Cell Biology." *Int J Cell Biol*, 563818. doi: 10.1155/2015/563818.

Danishefsky, I. & Bella, A. Jr. (1966). "The sulfated mucopolysaccharides from human umbilical cord." *J Biol Chem*, 241 (1), 143-6.

Dawson, M. H. (1937). "A serologically inactive polysaccharide elaborated by mucoid strains of group a hemolytic streptococcus." *J. Biol. Chem.* doi.org/10.1016/S0021-9258(18)74517-1)

Dicker, K. T., Gurski, L. A., Pradhan-Bhatt, S., Witt, R. L., Farach-Carson, M. C. & Jia, X. (2014). "Hyaluronan: a simple polysaccharide with diverse biological functions." *Acta Biomater*, 10 (4), 1558-70. doi: 10.1016/j.actbio.2013.12.019.

Frantz, C., Stewart, K. M. & Weaver, V. M. (2010). "The extracellular matrix at a glance." *J Cell Sci*, 123 (Pt 24), 4195-200. doi: 10.1242/jcs.023820.

Fraser, J. R., Laurent, T. C. & Laurent, U. B. (1997). "Hyaluronan: its nature, distribution, functions and turnover." *J Intern Med*, 242 (1), 27-33. doi: 10.1046/j.1365-2796.1997.00170.x.

Furnus, C. C., Valcarcel, A., Dulout, F. N., et al. (2003). "The hyaluronic acid receptor (CD44) is expressed in bovine oocytes and early stage embryos." *Theriogenology*, 60 (9), 1633-44. doi: 10.1016/s0093-691x(03)00116-x.

Garantziotis, S. & Savani, R. C. (2019). "Hyaluronan biology: A complex balancing act of structure, function, location and context." *Matrix Biol*, 78-79, 1-10. doi: 10.1016/j.matbio.2019.02.002.

Garantziotis, S., Li, Z., Potts, E. N., Kimata, K., Zhuo, L., Morgan, D. L., Savani, R. C., Noble, P. W., Foster, W. M., Schwartz, D. A. & Hollingsworth, J. W. (2009). "Hyaluronan mediates ozone-induced airway hyperresponsiveness in mice." *J Biol Chem*, 284 (17), 11309-17. doi: 10.1074/jbc.M802400200.

Girish, K. S. & Kemparaju, K. (2007). "The magic glue hyaluronan and its eraser hyaluronidase: a biological overview." *Life Sci*, 80 (21), 1921-43. doi: 10.1016/j.lfs.2007.02.037.

Hamerman, D. & Schuster, H. (1958). "Hyaluronate in normal human synovial fluid." *J Clin Invest*, 37 (1), 57-64. doi: 10.1172/JCI103585.

Hardwick, C., Hoare, K., Owens, R., Hohn, H. P., Hook, M., Moore, D., Cripps, V., Austen, L., Nance, D. M. & Turley, E. A. (1992). "Molecular cloning of a novel hyaluronan receptor that mediates tumor cell motility." *J Cell Biol*, 117 (6), 1343-50. doi: 10.1083/jcb.117.6.1343.

Harris, E. N., Kyosseva, S. V., Weigel, J. A. & Weigel, P. H. (2007). "Expression, processing, and glycosaminoglycan binding activity of the recombinant human 315-kDa hyaluronic acid receptor for endocytosis (HARE)." *J Biol Chem*, 282 (5), 2785-97. doi: 10.1074/jbc.M607787200.

Heldin, P., Lin, C. Y., Kolliopoulos, C., Chen, Y. H. & Skandalis, S. S. (2019). "Regulation of hyaluronan biosynthesis and clinical impact of excessive hyaluronan production." *Matrix Biol*, 78-79, 100-117. doi: 10.1016/j.matbio.2018.01.017.

Ialenti, A. & Di Rosa, M. (1994). "Hyaluronic acid modulates acute and chronic inflammation." *Agents Actions*, 43 (1-2), 44-7. doi: 10.1007/BF02005763.

Itano, N., Sawai, T., Yoshida, M., Lenas, P., Yamada, Y., Imagawa, M., Shinomura, T., Hamaguchi, M., Yoshida, Y., Ohnuki, Y., Miyauchi, S., Spicer, A. P., McDonald, J. A. & Kimata, K. (1999). "Three isoforms of mammalian hyaluronan synthases have distinct enzymatic properties." *J Biol Chem*, 274 (35), 25085-92. doi: 10.1074/jbc.274.35.25085.

Jentsch, H., Pomowski, R., Kundt, G. & Gocke, R. (2003). "Treatment of gingivitis with hyaluronan." *J Clin Periodontol*, 30 (2), 159-64. doi: 10.1034/j.1600-051x.2003.300203.x.

Jiang, D., Liang, J. & Noble, P. W. (2011). "Hyaluronan as an immune regulator in human diseases." *Physiol Rev*, 91 (1), 221-64. doi: 10.1152/physrev.00052.2009.

Jiang, D., Liang, J., Fan, J., Yu, S., Chen, S., Luo, Y., Prestwich, G. D., Mascarenhas, M. M., Garg, H. G., Quinn, D. A., Homer, R. J., Goldstein, D. R., Bucala, R., Lee, P. J., Medzhitov, R. & Noble, P. W. (2005). "Regulation of lung injury and repair by Toll-like receptors and hyaluronan." *Nat Med*, 11 (11), 1173-9. doi: 10.1038/nm1315.

Juhlin, L. (1997). "Hyaluronan in skin." *J Intern Med*, 242 (1), 61-6. doi: 10.1046/j.1365-2796.1997.00175.x.

Kaneiwa, T., Mizumoto, S., Sugahara, K. & Yamada, S. (2010). "Identification of human hyaluronidase-4 as a novel chondroitin sulfate hydrolase that preferentially cleaves the galactosaminidic linkage in the trisulfated tetrasaccharide sequence." *Glycobiology*, 20 (3), 300-9. doi: 10.1093/glycob/cwp174.

Kaye, M. A. & Stacey, M. (1950). "Observations on the chemistry of hyaluronic acid." *Biochem J*, 2, 13.

Kimura, M., Maeshima, T., Kubota, T., Kurihara, H., Masuda, Y. & Nomura, Y. (2016). "Absorption of Orally Administered Hyaluronan." *J Med Food*, 19 (12), 1172-1179. doi: 10.1089/jmf.2016.3725.

Kobayashi, T., Chanmee, T. & Itano, N. (2020). "Hyaluronan: Metabolism and Function." *Biomolecules*, 10 (11). doi: 10.3390/biom10111525.

Kothapalli, D., Flowers, J., Xu, T., Pure, E. & Assoian, R. K. (2008). "Differential activation of ERK and Rac mediates the proliferative and anti-proliferative effects of hyaluronan and CD44." *J Biol Chem*, 283 (46), 31823-9. doi: 10.1074/jbc.M802934200.

Linden, S. K., Sutton, P., Karlsson, N. G., Korolik, V. & McGuckin, M. A. (2008). "Mucins in the mucosal barrier to infection." *Mucosal Immunol*, 1 (3), 183-97. doi: 10.1038/mi.2008.5.

Liu, Y. Y., Lee, C. H., Dedaj, R., Zhao, H., Mrabat, H., Sheidlin, A., Syrkina, O., Huang, P. M., Garg, H. G., Hales, C. A. & Quinn, D. A. (2008). "High-molecular-weight hyaluronan a possible new treatment for sepsis-induced lung injury: a preclinical study in mechanically ventilated rats." *Crit Care*, 12 (4), R102. doi: 10.1186/cc6982.

Maleki, Atoosa, Anna-Lena Kjøniksen. & Bo Nyström. (2008). "Effect of pH on the Behavior of Hyaluronic Acid in Dilute and Semidilute Aqueous Solutions." *Macromolecular Symposia*, 274 (1), 131-140. doi: 10.1002/masy.200851418.

Martin-Deleon, P. A. (2011). "Germ-cell hyaluronidases: their roles in sperm function." *Int J Androl*, 34 (5 Pt 2), e306-18. doi: 10.1111/j.1365-2605.2010.01138.x.

Meyer, K. & Palmer, J. W. (1934). "The polysaccharide of the vitreous humor." *Journal of Biological Chemistry*.

Migliore, A., Massafra, U., Frediani, B., Bizzi, E., Sinelnikov Yzchaki, E., Gigliucci, G., Cassol, M. & Tormenta, S. (2017). "HyalOne(R) in the treatment of symptomatic hip OA - data from the ANTIAGE register: seven years of observation." *Eur Rev Med Pharmacol Sci*, 21 (7), 1635-1644.

Misra, S., Hascall, V. C., Markwald, R. R. & Ghatak, S. (2015). "Interactions between Hyaluronan and Its Receptors (CD44, RHAMM) Regulate the Activities of Inflammation and Cancer." *Front Immunol*, 6, 201. doi: 10.3389/fimmu.2015.00201.

Moon, K. C., Kim, K. B., Han, S. K., Jeong, S. H. & Dhong, E. S. (2019). "Assessment of Long-term Outcomes of Soft-Tissue Augmentation by Injecting Fibroblasts Suspended in Hyaluronic Acid Filler." *JAMA Facial Plast Surg*, 21 (4), 312-318. doi: 10.1001/jamafacial.2018.2127.

Mouta Carreira, C., Nasser, S. M., di Tomaso, E., Padera, T. P., Boucher, Y., Tomarev, S. I. & Jain, R. K. (2001). "LYVE-1 is not restricted to the lymph vessels: expression in normal liver blood sinusoids and down-regulation in human liver cancer and cirrhosis." *Cancer Res.*, 61 ((22)), 8079-84.

Neal, M. D., Leaphart, C., Levy, R., Prince, J., Billiar, T. R., Watkins, S., Li, J., Cetin, S., Ford, H., Schreiber, A. & Hackam, D. J. (2006). "Enterocyte TLR4 mediates phagocytosis and translocation of bacteria across the intestinal barrier." *J Immunol*, 176 (5), 3070-9. doi: 10.4049/jimmunol.176.5.3070.

Neuman, M. G., Nanau, R. M., Oruna-Sanchez, L. & Coto, G. (2015). "Hyaluronic acid and wound healing." *J Pharm Pharm Sci*, 18 (1), 53-60. doi: 10.18433/j3k89d.

Nolan, A., Baillie, C., Badminton, J., Rudralingham, M. & Seymour, R. A. (2006). "The efficacy of topical hyaluronic acid in the management of recurrent aphthous ulceration." *J Oral Pathol Med*, 35 (8), 461-5. doi: 10.1111/j.1600-0714.2006.00433.x.

Oe, M., Tashiro, T., Yoshida, H., Nishiyama, H., Masuda, Y., Maruyama, K., Koikeda, T., Maruya, R. & Fukui, N. (2016). "Oral hyaluronan relieves knee pain: a review." *Nutr J*, 15, 11. doi: 10.1186/s12937-016-0128-2.

Ogston, A. G. & Phelps, C. F. (1961). "The partition of solutes between buffer solutions and solutions containing hyaluronic acid." *Biochem J*, 78, 827-33. doi: 10.1042/bj0780827.

Ogston, A. G. & Sherman, T. F. (1961). "Effects of hyaluronic acid upon diffusion of solutes and flow of solvent." *J Physiol*, 156, 67-74. doi: 10.1113/jphysiol.1961.sp006658.

Ogston, A. G. & Phelps, C. F. (1961). "The partition of solutes between buffer solutions and solutions containing hyaluronic acid." *Biochem J*, 78, 827-33. doi: 10.1042/bj0780827.

Pavicic, T., Gauglitz, G. G., Lersch, P., Schwach-Abdellaoui, K., Malle, B., Korting, H. C. & Farwick, M. (2011). "Efficacy of cream-based novel formulations of hyaluronic acid of different molecular weights in anti-wrinkle treatment." *J Drugs Dermatol*, 10 (9), 990-1000.

Perry, K., Haresign, W., Wathes, D. C., et al. (2010). "Hyaluronan (HA) content, the ratio of HA fragments and the expression of CD44 in the ovine cervix vary with the stage of the oestrous cycle. Reproduction." *Reproduction*. doi: 10.1530/REP-09-0424. PubMed PMID: 20413624.

Ponta, H., Sherman, L. & Herrlich, P. A. (2003). "CD44: from adhesion molecules to signalling regulators." *Nat Rev Mol Cell Biol*, 4 (1), 33-45. doi: 10.1038/nrm1004.

Ponta, Helmut, David Wainwright. & Peter Herrlich. (1998). "Molecules in focus The CD44 protein family." *The International Journal of Biochemistry & Cell Biology*, 30 (3), 299-305. doi: 10.1016/s1357-2725(97)00152-0.

Prasetyo, A. D., Prager, W., Rubin, M. G., Moretti, E. A. & Nikolis, A. (2016). "Hyaluronic acid fillers with cohesive polydensified matrix for soft-tissue augmentation and rejuvenation: a literature review." *Clin Cosmet Investig Dermatol*, 9, 257-80. doi: 10.2147/CCID. S106551.

Prevo, R., Banerji, S., Ferguson, D. J., Clasper, S. & Jackson, D. G. (2001). "Mouse LYVE-1 is an endocytic receptor for hyaluronan in lymphatic endothelium." *J Biol Chem*, 276 (22), 19420-30. doi: 10.1074/jbc.M011004200.

Rodriguez-Merchan, E. C. (2013). "Intra-articular Injections of Hyaluronic Acid and Other Drugs in the Knee Joint." *HSS J*, 9 (2), 180-2. doi: 10.1007/s11420-012-9320-x.

Rubas, Werner. & George M. Grass. (1991). "Gastrointestinal lymphatic absorption of peptides and proteins." *Advanced Drug Delivery Reviews*, 7 (1), 15-69. doi: 10.1016/0169-409x(91)90047-g.

Sahana, T. G. & Rekha, P. D. (2018). "Biopolymers: Applications in wound healing and skin tissue engineering." *Mol Biol Rep*, 45 (6), 2857-2867. doi: 10.1007/s11033-018-4296-3.

Soltes, L., Mendichi, R., Kogan, G., Schiller, J., Stankovska, M. & Arnhold, J. (2006). "Degradative action of reactive oxygen species on hyaluronan." *Biomacromolecules*, 7 (3), 659-68. doi: 10.1021/bm050867v.

Spicer, A. P. & McDonald, J. A. (1998). "Characterization and molecular evolution of a vertebrate hyaluronan synthase gene family." *J Biol Chem*, 273 (4), 1923-32. doi: 10.1074/jbc.273.4.1923.

Takeuchi, Osamu, Katsuaki Hoshino, Taro Kawai, Hideki Sanjo, Haruhiko Takada, Tomohiko Ogawa, Kiyoshi Takeda. & Shizuo Akira. (1999). "Differential Roles of TLR2 and TLR4 in Recognition of Gram-Negative and Gram-Positive Bacterial Cell Wall Components." *Immunity*, 11 (4), 443-451. doi: 10.1016/s1074-7613(00)80119-3.

Tamer, T. M. (2013). "Hyaluronan and synovial joint: function, distribution and healing." *Interdiscip Toxicol*, 6 (3), 111-25. doi: 10.2478/intox-2013-0019.

Tammi, M. I., Day, A. J. & Turley, E. A. (2002). "Hyaluronan and homeostasis: a balancing act." *J Biol Chem*, 277 (7), 4581-4. doi: 10.1074/jbc.R100037200.

Tien, J. Y. & Spicer, A. P. (2005). "Three vertebrate hyaluronan synthases are expressed during mouse development in distinct spatial and temporal patterns." *Dev Dyn*, 233 (1), 130-41. doi: 10.1002/dvdy.20328.

Toole, B. P. (2001). "Hyaluronan in morphogenesis." *Semin Cell Dev Biol*, 12 (2), 79-87. doi: 10.1006/scdb.2000.0244.

Toole, B. P. (2004). "Hyaluronan: from extracellular glue to pericellular cue." *Nat Rev Cancer*, 4 (7), 528-39. doi: 10.1038/nrc1391.

Triggs-Raine, B., Salo, T. J., Zhang, H., Wicklow, B. A. & Natowicz, M. R. (1999). "Mutations in HYAL1, a member of a tandemly distributed multigene family encoding disparate hyaluronidase activities, cause a newly described lysosomal disorder, mucopolysaccharidosis IX." *Proc Natl Acad Sci U S A*, 96 (11), 6296-300. doi: 10.1073/pnas.96.11.6296.

Turley, E. A. & Asculai, S. S. (2003). "*Oral administration of effective amounts of forms of hyaluronic acid.*" USA, US6537978B1.

Turley, E. A., Noble, P. W. & Bourguignon, L. Y. (2002). "Signalling properties of hyaluronan receptors." *J Biol Chem*, 277 (7), 4589-92. doi: 10.1074/jbc.R100038200.

Tzircotis, G., Thorne, R. F. & Isacke, C. M. (2005). "Chemotaxis towards hyaluronan is dependent on CD44 expression and modulated by cell type variation in CD44-hyaluronan binding." *J Cell Sci*, 118 (Pt 21), 5119-28. doi: 10.1242/jcs.02629.

Vigetti, D., Viola, M., Karousou, E., De Luca, G. & Passi, A. (2014). "Metabolic control of hyaluronan synthases." *Matrix Biol*, 35, 8-13. doi: 10.1016/j.matbio.2013.10.002.

Voigt, J. & Driver, V. R. (2012). "Hyaluronic acid derivatives and their healing effect on burns, epithelial surgical wounds, and chronic wounds: a systematic review and meta-analysis of randomized controlled trials." *Wound Repair Regen*, 20 (3), 317-31. doi: 10.1111/j.1524-475X.2012.00777.x.

Weissmann, B. & Meyer, K. (1954). "The structure of hyalobiuronic acid and of hyaluronic acid from umbilical cord." *J. Am. Chem. Soc.*, 76, 1753–1757.

Wróbel, T., Dziegiel, P., Mazur, G., Zabel, M., Kuliczkowski, K. & Szuba, A. (2005). "LYVE-1 expression on high endothelial venules (HEVs) of lymph nodes. Lymphology." *Lymphology*, 38 ((3)), 107-10.

Zhou, B., Weigel, J. A., Saxena, A. & Weigel, P. H. (2002). "Molecular cloning and functional expression of the rat 175-kDa hyaluronan receptor for endocytosis." *Mol Biol Cell*, 13 (8), 2853-68. doi: 10.1091/mbc.02-03-0048.

In: Hyaluronic Acid
Editor: Vittorio Unfer

ISBN: 978-1-53619-743-3
© 2021 Nova Science Publishers, Inc.

Chapter 2

HYALURONIC ACID IN OBSTETRICS: ROLE IN PHYSIOLOGICAL PREGNANCY

Fabio Facchinetti[1] and Vittorio Unfer[2,*]
[1]Unit of Obstetrics and Gynecology, Mother-Infant Department, University of Modena and Reggio Emilia, Modena, Italy
[2]Systems Biology group lab Rome Italy

ABSTRACT

Pregnancy is a physiologically fine-balanced condition, characterized by radical hormonal and physical changes necessary for successful blastocyst implantation, optimal embryo development and safe birth. During gestation, several biological modifications are adopted by maternal body to accept the fetus (e.g., immune tolerance). In this context, Hyaluronic Acid (HA) is an interesting molecule, seemingly involved in many steps of the process, with a pivotal role in ovulation, fertilization, blastocyst implantation, inhibition of uterine contraction, immunomodulation of T cells or labour-related cervical modifications.

Keywords: HMW-HA, competent blastocyst, receptive uterus, immune tolerance, progesterone, PGRMC1

---

* Corresponding Author's Email: vunfer@gmail.com.

1. Physiological Role of HA from Ovulation to Labour

1.1. HA Activities during Ovulation

Hyaluronan is a fundamental component of the extracellular matrix (ECM) that surrounds ovarian follicles (Irving-Rodgers and Rodgers 2005). HA, besides being a structural element, is a key factor that leads to oocyte maturation and ovulation (Rodgers, Irving-Rodgers, and Russell 2003, N Kimura 2007). Produced by granulosa and cumulus cells (Salustri et al. 1992), HMW-HA was detected also in follicular fluid, where it creates an osmotic gradient responsible for follicular fluid accumulation and antrum formation (Clarke et al. 2006, Rodgers and Irving-Rodgers 2010).

Many authors have reported that reproductive hormones can differentially regulate the expression of Hyaluronan Synthase (Has) enzymes as Has2 (the main HA synthase found in human ovaries), Has3 and HA receptors (as CD44) (Chavoshinejad et al. 2016). Oestradiol, together with insulin and Follicle Stimulating Hormone (FSH), stimulates Has2 expression during cumulus cell expansion prior to ovulation. Moreover, oestradiol and luteinizing hormone (LH) increase Has3 and CD44 mRNA expression in the granulosa cells during ovulation. High expression of Has3 and CD44 was also detected in the corpus luteum, indicating a pattern of expression in the ovaries during the oestrous cycle.

These evidence suggest that HMW-HA synthesis occurs during follicular maturation and cumulus cell expansion, meanwhile LMW-HA synthesis is predominant after LH surge. LMW-HA has a major role during inflammatory and angiogenesis (Collins et al. 2011, Rayahin et al. 2015), characteristic of the follicles during ovulation (Blundell et al. 2003) and corpus luteum formation (Berisha B 2016, Skarzynski et al. 2013). On the contrary, HMW-HA is mainly involved in oocyte maturation.

In mammals, the oocyte is surrounded by multiple layers of compacted cumulus cells that form the cumulus-oocyte complex (COC). During oocyte maturation, COC drastically expands, protecting the oocyte from the proteolytic and mechanical stress occurring during ovulation (Magier et al. 1990) and supporting its fertilisation (Tanghe et al. 2002, Russell et al. 2016). Experimental studies evidenced that HA has a pivotal role in cumulus expansion in the COCs (Nagyova 2012). In *in-vitro* studies, HAS inhibitors failed cumulus expansion (Yokoo, Kimura, and Sato 2010), thus highlighting that hyaluronan synthesis and its accumulation is crucial for cumulus expansion of COCs (Yokoo and Sato 2011). In this regard, Has2 and CD44 expression in cumulus cells can be considered as a marker of oocyte competence, while CD44 in the follicular fluid can be associated with good quality oocytes (Ohta et al. 2001, Assidi et al. 2008). Moreover, CD44 expression was found in COCs during oocyte maturation (Kimura et al. 2002). When COCs were cultured with anti-CD44 antibody, oocyte maturation was inhibited in an antibody concentration-dependent manner, which had a blocking effect on the hyaluronan-binding ability of CD44. These results suggest that the binding between hyaluronan and CD44 during cumulus expansion is important for oocyte maturation (Yokoo and Sato 2011) as it regulates tyrosine phosphorylation of Connexin 43, the main gap junction protein in COCs. Interaction between HA and CD44 leads to the closure of the gap junctions and the activation of the maturation promotion factor (MPF), a protein involved in oocyte meiosis resumption (Sato E 2005).

2. EARLY STAGES OF ZYGOTE DEVELOPMENT AND IMPLANTATION

Several studies reported that supplementation of media culture with HMW-HA improves blastocyst adhesion, embryo development and viability both in human and animal models (Fouladi-Nashta et al. 2017, Urman et al. 2008, Furnus, de Matos, and Martinez 1998, Gardner, Rodriegez-Martinez, and Lane 1999, Lane et al. 2003, Jang et al. 2003, Toyokawa, Harayama, and Miyake 2005), increasing also cryotolerance of blastocysts (Lane et al. 2003, Dattena et al. 2007).

Furthermore, Marei et al. (Marei, Salavati, and Fouladi-Nashta 2013) demonstrated that the inhibition of HA synthesis with 4-

methylumbelliferone (4-MU) induces the suppression of blastocyst development. This effect is reversible after removing 4-MU from the culture medium, confirming the critical role of HA in the development of the embryo (Nakamura et al. 2004). In fact, 4-MU inhibits the synthesis of HA because it binds to HAS substrates and therefore downregulates the expression of HAS2 and HAS3 (Kultti et al. 2009).

Both HMW-HA and LMW-HA have an important role for embryo development, even if they have different functions. LMW-HA is mainly a pro-angiogenic factor that acts in the early stages of zygote development (until morula) and during the attachment (Marei, Salavati, and Fouladi-Nashta 2013); HMW-HA is mainly expressed during implantation, supporting blastocyst adhesion, and during the entire pregnancy, maintaining the immune-tolerance state until labour (Hadas et al. 2020, Petrey and de la Motte 2014, Jiang, Liang, and Noble 2011, McKeown-Longo and Higgins 2021). This is also confirmed by the differential expression of HA enzymes and receptors during embryo development, thus suggesting the presence of a decreasing HA molecular size gradient. Both HAS2 and HAS3 transcripts were found in the oviducts of *in-vivo* models (Mohey-Elsaeed et al. 2016, Tienthai et al. 2003, Ulbrich et al. 2004). Has3, responsible for LMW-HA synthesis, is less expressed in the ampulla compared to the rest of the oviduct (Mohey-Elsaeed et al. 2016, Ulbrich et al. 2004). Moreover, LMW-HA regulates the expression of insulin-like growth factors (IGFs) (Homandberg, Ummadi, and Kang 2004) and heat shock proteins (Xu et al. 2002), which are important for early embryo development in the oviduct (Aviles, Gutierrez-Adan, and Coy 2010). Interestingly, Hyal2 mRNA is expressed in embryos until the morula stage and in the oviduct, with significantly higher levels in the isthmus than in the ampulla (Marei, Salavati, and Fouladi-Nashta 2013). LMW-HA supports embryo development until the morula stage, since embryos in the cleavage stage do not express Hyal2 (Marei, Salavati, and Fouladi-Nashta 2013).

During morula stage, LMW-HA increases cell proliferation by binding CD44 and RHAMM receptors (Xu et al. 2002), thus activating MAPK pathway (Zhu et al. 2013) and stimulating mitosis. Under *in-vitro* conditions, supplementation of HYAL2 is beneficial for preimplantation embryo development and embryo quality. In fact, the degradation of HA by HYAL2 into smaller fragments is known to activate CD44 signalling (Ohno-Nakahara et al. 2004), which leads to the stimulation of the MAPK pathway required for embryo development. These effects are abrogated by adding an anti-CD44 antibody or a MAPK inhibitor (Marei, Salavati, and Fouladi-Nashta 2013).

After reaching the implantation site, the embryo attaches to the endometrial surface of the uterus and invades the epithelium and the maternal circulation to form the placenta. (Kim and Kim 2017). For a successful implantation, competent blastocyst, receptive uterus and physiological oestrogen levels are fundamental (Bedzhov and Zernicka-Goetz 2014, Hadas et al. 2020). Blastocyst attachment to the endometrial epithelium is mediated by ECM (Cross, Werb, and Fisher 1994). In this regard, HMW-HA physiologically facilitates attachment of the embryo before implantation, interacting with CD44 and upregulating osteopontin, another ligand of CD44 upregulated by progesterone (Hadas et al. 2020, McKeown-Longo and Higgins 2021, Berneau et al. 2019). Expression of Has2 and HA receptors (TLR-4 and CD44 at the attachment interface; LYVE-1 and RHAMM in the underlying stroma) and their accumulation at the uterine-blastocyst interface confirm that HMW-HA have both a structural and a regulatory role in this process (Hadas et al. 2020, Berneau et al. 2019).

In the final stages before implantation, the blastocyst undergoes lineage differentiation, generating the polar and the mural trophectoderm, in which Has1 and Has2 are highly expressed, thus confirming HA accumulation during the implantation phase. Similarly, CD44 was observed in blastocysts and in the uterine luminal epithelium (Hadas et al. 2020). HAS1 or HAS2 knock-out mice (both single and double mutant) undergo a failure in blastocyst adhesion.

Accumulation of HA, the expression of its receptors, and the impact of HA synthases deletion, confirm the involvement of HMW-HA in implantation and a successful attachment and primary decidualization (Hadas et al. 2020).

3. HA Importance for Decidual Cells

Decidualization is a morphological and functional process, in which the endometrium undergoes a significant change to form the decidual lining into which the blastocyst implants. During the decidual reaction, Endometrial Stromal Cells (ESCs) differentiate into Decidual Stromal Cells (DSCs) through a genetic reprogramming in which genes involved in the pro-inflammatory response are down-regulated and genes involved in cellular proliferation, tolerance and tissue invasion are up-regulated. Decidual Stromal Cells (DSCs) are the major cellular component of maternal–fetal interface and represent 80% of decidual cells. Despite the importance of DSCs at the maternal-fetal interface, their biological role is still not completely defined, but differentiation of ESCs into DSCs represents a critical moment for embryo implantation and pregnancy establishment (Ledee et al. 2011, Xu et al. 2012).

The major driver of decidualization is progesterone, produced by the corpus luteum of the ovary following ovulation. In absence of pregnancy, the corpus luteum regresses in 14 days, progesterone is downregulated and menstruation occurs. On the contrary, in presence of pregnancy, the production of human chorionic gonadotropin (hCG) by trophoblast maintains progesterone levels, which in turn modulates maternal immune response by suppressing inflammatory responses, reduces uterine contractility and improves utero-placental circulation and luteal phase support (Ng et al. 2020).

In addition to their traditional nutrition and support to embryo in pregnancy, growing evidence supports the idea that DSCs are involved in immune regulatory process, such as the production of cytokines and antigen phagocytosis (Olivares et al. 1997, Ruiz C 1997, Piao et al. 2012). They play a crucial role in the regulation of decidual $CD4^+$T-cell cytokine production, which helps to maintain a balanced cytokine concentration at the maternal/fetal interface.

DSCs are involved in the production of ECM components (Lockwood et al. 2009, Godbole et al. 2011, Du et al. 2012) by mediating extravillous trophoblast (EVT) invasion and homeostatic protection, and by acting as embryo quality sensor upon implantation (Teklenburg et al. 2010).

Both HA and CD44 have been observed in early human embryos and in the decidual stroma (Marzioni et al. 2001, Goshen et al. 1996). Moreover it was demonstrated that hyaluronan-enriched transfer medium significantly increases pregnancy and implantation rates in patients with multiple embryo transfer failures, suggesting that HA is essential for embryo implantation and pregnancy (Nakagawa et al. 2012).

The importance of HMW-HA and CD44 in pregnancy is confirmed by the high concentration of HMW-HA secreted by DSCs, and by the CD44 expression on the membranes of these cells (Zhu et al. 2013). HMW-HA promotes proliferation and inhibits apoptosis of DSCs in CD44-dependent manner via PI3K/Akt and MAPK/ERK1/2 signalling pathways. Moreover, HA content, HA molecular weight, HAS2 mRNA level, and CD44 expression were significantly decreased in women with unexplained miscarriage (Zhu et al. 2013) compared with normal pregnancy, thus indicating that higher level and greater molecular mass of HA at maternal-fetal interface contributes to DSCc growth and their maintenance in human early pregnancy.

DSCs are the main ECM producer in the decidua, and they are influenced by matrix environment and sex hormones, which are fundamental at the maternal–fetal interface where they regulate HMW-HA production. During synthesis, growing polymers of HMW-HA are released through the membrane into the extracellular space, where they stimulate DSCs survival and counteract apoptosis in an autocrine manner. Several hormones associated to pregnancy are involved in this process, including oestrogen, progesterone, and human chorionic gonadotropin (hCG), which stimulate HAS2 transcription and HA production by DSCs (Patriarca et al. 2013, Stock et al. 2002).

4. IMMUNOMODULATION BY HMW-HA

Pregnancy is a natural process that poses an immunological challenge because non-self-fetus must be accepted. Scientific evidence showed that during a successful pregnancy, the maternal-fetal interface is an immunologically active area with a series of immune processes that maintain a fine balance between maternal-fetal tolerance and anti-pathogens defense. This balance is mainly established locally, where the extravillous trophoblast cells invade the decidua and come into contact with the decidual immune cells (DICs). Failure or alterations in the crosstalk between these cells is associated with complications of pregnancy, such as preeclampsia and recurrent spontaneous abortion (RSA) (Arck and Hecher 2013).

Considering the importance of immune tolerance for successful pregnancy, the anti-inflammatory activities of HMW-HA are fundamental during gestation. Beyond the physical properties as a structural component of ECM, it has become clear that HMW-HA functions as tissue integrity signal and suppresses inflammatory response. In particular HMW-HA (1) induces the secretion of anti-inflammatory cytokines such as IL-10 in decidual macrophages (Wang et al. 2019); (2) inhibits the expression of proinflammatory factors as TNF-α or IFN-γ (Wang et al. 2019); (3) stimulates immunomodulation by cell differentiation e.g., T-naïve into Treg (Bollyky et al. 2011), and decidual macrophages into M2 at the maternal–fetal interface by CD44 activation of PI3K/Akt-STAT-3/STAT-6 signalling pathways (Wang et al. 2019); (4) induces the expression of progesterone receptors (Zhao et al. 2014); (5) blocks monocytes and pro-inflammatory factors in chronic inflammatory diseases (Petrey and de la Motte 2014); (6) shields TLRs 2/4 receptors, inhibiting their stimulation. In particular, the latter represents the perfect example that shows the antithetical properties of HA with different molecular weights. Indeed, HMW-HA creates a dense and viscous protective coat around the cells, thus shielding surface receptors such as TLRs. As a consequence, their interactions with ligands are limited, reducing pro-inflammatory stimulations. On the other hand, LMW-HA directly interacts with TLRs, inducing a proinflammatory state. LMW-HA is typical in damaged, injured or infected tissues, subjected to inflammation or oxidative stress. In these cases, LMW-HA represents a defensive mechanism that stimulates innate immune response (Fallacara et al. 2018).

5. HYALURONIC ACID INVOLVEMENT IN CERVICAL MODIFICATIONS

HMW-HA plays a fundamental role in female reproductive biology, from folliculogenesis to birth, and it is also responsible for cervical remodeling process (Skinner SJ 1981, DN. 1947). Cervical structure protects the foetus in the uterus for 37-40 weeks of human gestation, until it is ready for extra-uterine life (Mahendroo 2019). At the end of pregnancy, the cervix must be able to widen to 10 cm to allow birth and it is therefore necessary that tissues undergo a gradual remodeling. This process can be divided into 3 overlapping phases: softening, ripening and dilation (Nallasamy and Mahendroo 2017). In particular, increased synthesis of HA was reported during cervical ripening by several authors (Golichowski, King, and Mascaro 1980, Danforth et al. 1974).

Uterine cervical ripening is characterized by remodeling of extracellular matrix. Its components, collagen and glycosaminoglycans (GAGs), change quantitatively and qualitatively during this process. It is established that, at the onset of labour, the HA content of the cervix markedly increases, as evidenced also in *in-vivo* model. Indeed, during cervical ripening, the total GAG content doubles and 71% of the total GAGs are represented by HA in pregnant mice, compared to 51% in the non-pregnant animals. Moreover the HA content in mouse cervix increases from 7.4 nmol/mg in non-pregnant mice to 24.6 nmol/mg in pregnant ones (Mahendroo 2019).

Due to its chemo-physical properties, HA contributes first to the uterine cervical softening, maintaining tissue viscoelasticity and increasing its hydration; second, it promotes degradation of collagen in the stroma during cervical ripening. Has2, the main hyaluronan synthase (HAS), is involved in cervical ripening, and it is responsible for the increased cervical HA synthesis. Little is known about the factors that regulate Has2 transcription in the cervix, but several studies have confirmed that estrogens, progesterone and the peptide hormone relaxin are its regulatory molecules (Mahendroo 2019). In this regard Akgul et al. found that Has2 expression was reduced after the administration of an estrogen receptor antagonist, while it was increased in the presence of exogenous oestrogens.

The overexpression of hyaluronidases activity during cervical ripening is fundamental in human and mouse cervix (Straach et al. 2005) to convert HMW-HA to LMW-HA fragments postpartum (Mahendroo 2019), promoting the synthesis of anti-microbial peptides and inducing wound healing by increasing the migratory capacity of epithelia (Dusio et al. 2011).

The presence of HMW-HA in the cervix before labour is not only important for the ripening but also for the immune-regulatory activity. In this regard, TLR expression in the cervix is widely reported (Ruscheinsky, De la Motte, and Mahendroo 2008, Gonzalez et al. 2007).

LMW-HA is a ligand for two members of this receptor family, TLR2 and TLR4, similarly to pathogen-associated molecular elements such as lipopolysaccharides (LPS) (Gonzalez et al. 2007). Despite the TLR signalling pathway activated by LPS, the one activated by LMW-HA in vaginal epithelial cells does not induce Nuclear Factor-κB (NFκB) translocation in the nucleus.

The presence of LMW-HA also induces a relaxation of the tight junctions, promoting bacterial ascension along the cervix, which can trigger labour due to its proinflammatory activities. Additionally, several studies reported that HMW-HA depleted cervix was associated with a defective ability of HA to act as a barrier during pregnancy (Mahendroo 2019). Furthermore, the absence of HMW-HA caused epithelia disorders and facilitated bacterial ascension through the cervical canal (Fouladi-Nashta et al. 2017, Mahendroo 2019). Finally, HMW-HA is important because protects TLRs receptors, avoiding pro-inflammatory reactions, a fundamental activity to prevent preterm birth (PTB).

Prevention of Preterm Birth (PTB)

PTB, defined as delivery before the 37th week of gestation (Blencowe et al. 2013, Mahendroo 2019) remains one of the biggest unresolved obstetrical problems. It represents the leading cause of infant mortality in the first year, associated with an increased risk of neurological, respiratory and gastrointestinal complications throughout life. Prevention is complex because it is estimated that only 25-40% of PTB cases result from inflammation and/or infection, meanwhile the remaining 60% is of unknown aetiology (Ferrero et al. 2016). Most cases of infection-induced PTB are caused by bacterial ascension through the lower reproductive tract. Several authors report that the vaginal microbiome changes, and in particular it significantly decreases during pregnancy (Romero et al. 2014, DiGiulio et al. 2015). As noted above, HA deficiency in the cervix exposes the fetus to a greater risk of ascending infection and therefore to a greater risk of PTB (Akgul et al. 2014). Pathogen-induced PTB studies have identified a mechanism that induces loss of HA, thus helping to increase susceptibility to ascending infections. Group B Streptococcus (GBS) is a Gram-positive bacterium, present in the vaginal microbiome of 30% of women and it is well known to be a risk factor for PTB (Allen et al. 1999, Kolar SL 2015). Its transmission to the fetus can occur during vaginal birth or by ascending infection that leads to inflammation of the fetal membrane and therefore to PTB.

This bacterium bypasses the immune response and uses ECM factors to promote adherence, hemolytic activity to induce immune cell evasion, and hyaluronidase (HyalB) synthesis to degrade HMW-HA into disaccharide components that bind to TLR 2/4 receptors (Mahendroo 2019). Unlike LPS, LMW-HA and O-HA do not induce a specific immune response against GBS, thus facilitating bacterial survival (Kolar SL 2015). This happens because the binding of HA to TLRs induces a non-specific reaction against GBS by saturating receptors and preventing pathogen recognition. Vornhagen, J et al. showed a higher production of HyalB from GBS strains isolated from women in preterm labour compared to strains isolated from women with full-term pregnancy but not in labour (Vornhagen, Adams Waldorf, and Rajagopal 2017, Vornhagen et al. 2016). These data were also confirmed using a mouse model in which the GBS knockout strains of HYALB showed a reduced ability to ascend in the reproductive system.

This interesting finding correlates with a reduced PTB rate, confirming the virulence factor properties of HyalB (Mahendroo 2019).

The transition from a quiescent state of the uterus and cervix to a state of active labour depends on an inflammatory state mediated by estrogen and progesterone (Ackerman et al. 2016, Nadeem et al. 2016, Merlino et al. 2009, Kumar et al. 2015, Mesiano et al. 2002).

The role of progesterone in maintaining membrane homeostasis is unclear, as it is its involvement in the pro-labour inflammatory state. Inflammation plays a fundamental role in the homeostasis of the fetal membrane because it can induce their rupture (Lozovyy et al. 2021). The fetal membranes consist of two layers, amnion and chorion, separated by a mesenchymal layer of cells and ECM. The fetal amniochorion layer is a structure connected to the maternal decidua in a system called the fetal-maternal interface, whose main function is to protect the fetus during gestation until labour. Both the amnion and the chorion are subject to constant remodeling (Richardson, Taylor, and Menon 2020), and several studies have confirmed their progressive senescence during gestation until their rupture near birth in an inflammatory environment (Menon et al. 2014, Menon et al. 2013). This programmed senescence is mediated by p38MAPK, a key factor in oxidative stress (Dixon et al. 2018), which is prematurely activated in cases of preterm birth (PTB) and preterm premature rupture of the membranes (pPROM) (Menon et al. 2014).

Occasionally, progesterone treatment fails to reduce oxidative stress and/or p38MAPK activation, as well as to avoid senescence or inflammatory states in fetal membranes (Ayad, Taylor, and Menon 2018).

Such failure can be explained by its physiological silencing due to the downregulation of the PGRMC1 receptor (Ayad, Taylor, and Menon 2018). PGRMC1 is one of the major progesterone receptors in pregnancy and its expression is induced by HMW-HA (Zhao et al. 2014). In the presence of inflammatory states (e.g., the last phase of gestation), HMW-HA is subjected to degradation, leading to downregulation of PGRMC1 and accumulation of LMW-HA, which supports inflammation.

PGRMC1 and HA

Progesterone (P4) is one of the key regulators that control myometrial growth and contractility. The transition of the myometrium from a relaxed to a highly contractile stage is a critical event during parturition, and it is seemingly controlled by the combined effects of changes in P4 receptors and in oestrogen stimulation pathways. P4 production by the human placenta gradually increases with advancing pregnancy, from 200 to 600 mg daily at term (MacDonald, Dombroski, and Casey 1991), meanwhile, a functional P4 decrease seems to induce parturition (Mesiano 2001, Mitchell and Taggart 2009). P4 affects myometrial contractility through genomic and non-genomic pathways. Genomic actions of P4 are mediated by the classic nuclear P4 receptors (nPRs) that function as ligand-activated transcription factors and control the expression of specific genes regulating the contractile phenotype (Mesiano and Welsh 2007). Non-genomic pathways for P4, in contrast, are faster and are believed to directly affect contraction by modulating intracellular signal transduction pathways (Thomas 2008, Gellersen, Fernandes, and Brosens 2009) or by inhibiting calcium influx. Functional studies in PGR null female mice, as well as pharmacological studies using PGR antagonists (e.g., mifepristone), have demonstrated a fundamental role for PGR in female fertility (Conneely 2000). However, PGR does not appear to be the sole receptor mediating P4 actions, as cells that completely lack expression of PGR are still able to respond to P4 (Peluso et al. 2006, Peluso et al. 2009). Several studies have demonstrated non-classical P4 responses in the uterus. De Mayo and colleagues demonstrated that, while PGR is important for mediating changes in the expression of many P4-regulated genes in the murine uterus, many other genes are regulated by P4 through a PGR-independent mechanism. Two families of non-classical membrane receptors have been identified, the Progestin and AdipoQ receptor (PAQR) and the Progesterone Receptor Membrane Component (PGRMC) (Thomas 2008, Cahill 2007).

PGRMC1 and its close relative, PGRMC2, belong to the PGRMC family. PGRMC1 seems to be involved in a wide array of biological functions, including steroidogenesis and cellular homeostasis (Rohe et al. 2009). It is involved in fetal membrane integrity (Zhao et al. 2014) and patients with primary ovarian insufficiency (POI) have reduced levels of PGRMC1. Some studies demonstrated that PGRMC1 not only regulates cell viability and sexual steroid hormones biosynthesis, but it also mediates the antiapoptotic effects of P4 on granulosa cells (GCs), which provide the physical support and the microenvironment required for the developing oocyte (Peluso et al. 2010, Mansouri et al. 2008, Rohe et al. 2009, Peluso et al. 2006).

It is well recognized that changes in P4 levels or changes in its receptors may lead to the onset of labour (Mesiano and Welsh 2007). *In-vitro* contractility studies demonstrated that pretreatment of myometrial strips with PGRMC1 antibody suppresses the P4-induced relaxation, thus evidencing that PGRMC1 mediates the non-genomic action of P4 in inhibition the spontaneous contractility of human myometrium during pregnancy (Wu et al. 2011).

Thus, progesterone is fundamental for membrane homeostasis throughout pregnancy until parturition, and HA may support its activity.

Indeed, it has been recently demonstrated that HA, after binding to CD44 receptor, induces mRNA and PGRMC1 protein level up-regulation in GC tumor cell line and rat primary GCs (Zhao et al. 2014). Such up-regulation depends on epigenetic silencing of miR-139-5p, which directly targets PGRMC1 and it is responsible for apoptosis activation in GCs.

CONCLUSION

As widely reported, HA has a pivotal role in several phases of pregnancy, from fertilization to labour. Under physiological conditions, HA mainly exists as high molecular weight polymer, but it displays several regulatory activities and functional properties, based on its different molecular weight. Until now, clinical evidence widely suggested the use of LMW-HA in gynaecology for menopause-related symptoms, or in association with physical treatments (e.g., radiation therapy) to counteract the onset of adverse events. On the contrary, HMW-HA has been poorly evaluated as clinical treatment, but its physiological presence in the extracellular matrix of all the main organs and tissues involved in pregnancy (uterus, cervix, placenta, decidua, chorion, amnios, ovarium, etc.) suggests the importance of this molecule for a successful gestation.

HMW-HA represents a fundamental ECM component, which incorporates water molecules, reduces the transition of other small molecules and contributes to hinder the passage of viruses and bacteria through the pericellular zone. It modulates the immune response and acts as an anti-inflammatory molecule. This activity is fundamental for immune tolerance during pregnancy. The wide involvement of HMW-HA in all phases of gestation confirms its importance and supports its potential therapeutical application as a treatment for several gestational conditions connected with impaired HMW-HA metabolism. In fact, several papers shed light on its importance for blastocyst adhesion and implantation, for an efficient immune tolerance and for the correct development of the haemo-lymphatic system. HMW-HA is crucial during the whole gestation, resulting in regulatory activity on the PGRMC1 expression, a specific progesterone receptor expressed in maternal and fetal-maternal interface tissues, and involved in uterine quiescence. Reduced expression of PGRMC1 is frequent in patients with POI, a serious reproductive dysfunction in which the follicle pool is reduced or depleted. Notably, HMW-HA administration can restore PGRMC1 levels. Based on these data, oral treatment with HMW-HA may represent a promising opportunity to prevent recurrent spontaneous abortions and PTB in patients with risk factors. Moreover, the high safety profile of HMW-HA further supports its administration in pregnancy, thus leading to a successful and healthy gestation.

REFERENCES

Ackerman, W. E. th., Summerfield, T. L., Mesiano, S., Schatz, F., Lockwood, C. J. & Kniss, D. A. (2016). "Agonist-Dependent Downregulation of Progesterone Receptors in Human Cervical Stromal Fibroblasts." *Reprod Sci*, 23 (1), 112-23. doi: 10.1177/1933719115597787.

Akgul, Y., Word, R. A., Ensign, L. M., Yamaguchi, Y., Lydon, J., Hanes, J. & Mahendroo, M. (2014). "Hyaluronan in cervical epithelia protects against infection-mediated preterm birth." *J Clin Invest*, 124 (12), 5481-9. doi: 10.1172/JCI78765.

Allen, U., Nimrod, C., Macdonald, N., Toye, B., Stephens, D. & Marchessault, V. (1999). "Relationship between antenatal group B streptococcal vaginal colonization and premature labour." *Paediatr Child Health*, 4 (7), 465-9. doi: 10.1093/pch/4.7.465.

Arck, P. C. & Hecher, K. (2013). "Fetomaternal immune cross-talk and its consequences for maternal and offspring's health." *Nat Med*, 19 (5), 548-56. doi: 10.1038/nm.3160.

Assidi, M., Dufort, I., Ali, A., Hamel, M., Algriany, O., Dielemann, S. & Sirard, M. A. (2008). "Identification of potential markers of oocyte competence expressed in bovine cumulus cells matured with follicle-stimulating hormone and/or phorbol myristate acetate *in vitro*." *Biol Reprod*, 79 (2), 209-22. doi: 10.1095/biolreprod.108.067686.

Aviles, M., Gutierrez-Adan, A. & Coy, P. (2010). "Oviductal secretions: will they be key factors for the future ARTs?" *Mol Hum Reprod*, 16 (12), 896-906. doi: 10.1093/molehr/gaq056.

Ayad, M. T., Taylor, B. D. & Menon, R. (2018). "Regulation of p38 mitogen-activated kinase-mediated fetal membrane senescence by statins." *Am J Reprod Immunol*, 80 (4), e12999. doi: 10.1111/aji.12999.

Bedzhov, I. & Zernicka-Goetz, M. (2014). "Self-organizing properties of mouse pluripotent cells initiate morphogenesis upon implantation." *Cell*, 156 (5), 1032-44. doi: 10.1016/j.cell.2014.01.023.

Berisha, B., Schams, D., Rodler, D. & Pfaffl, M. W. (2016). "The Most Important Regulatory Event for Follicle and Corpus Luteum Development and Function in Cow - An Overview. ." *Anat Histol Embryol.*, 45 ((2)), 124-30. doi: doi:10.1111/ahe.

Berneau, S. C., Ruane, P. T., Brison, D. R., Kimber, S. J., Westwood, M. & Aplin, J. D. (2019). "Characterisation of Osteopontin in an *In Vitro* Model of Embryo Implantation." *Cells*, 8 (5). doi: 10.3390/cells8050432.

Blencowe, H., Cousens, S., Chou, D., Oestergaard, M., Say, L., Moller, A. B., Kinney, M., Lawn, J. & Group Born Too Soon Preterm Birth Action. (2013). "Born too soon: the global epidemiology of 15 million preterm births." *Reprod Health*, 10, Suppl 1, S2. doi: 10.1186/1742-4755-10-S1-S2.

Blundell, C. D., Mahoney, D. J., Almond, A., DeAngelis, P. L., Kahmann, J. D., Teriete, P., Pickford, A. R., Campbell, I. D. & Day, A. J. (2003). "The link module from ovulation- and inflammation-associated protein TSG-6 changes conformation on hyaluronan binding." *J Biol Chem*, 278 (49), 49261-70. doi: 10.1074/jbc. M309623200.

Bollyky, P. L., Wu, R. P., Falk, B. A., Lord, J. D., Long, S. A., Preisinger, A., Teng, B., Holt, G. E., Standifer, N. E., Braun, K. R., Xie, C. F., Samuels, P. L., Vernon, R. B., Gebe, J. A., Wight, T. N. & Nepom, G. T. (2011). "ECM components guide IL-10 producing regulatory T-cell (TR1) induction from effector memory T-cell precursors." *Proc Natl Acad Sci U S A*, 108 (19), 7938-43. doi: 10.1073/pnas.1017360108.

Cahill, M. A. (2007). "Progesterone receptor membrane component 1: an integrative review." *J Steroid Biochem Mol Biol*, 105 (1-5), 16-36. doi: 10.1016/j.jsbmb.2007.02.002.

Chavoshinejad, R., Marei, W. F., Hartshorne, G. M. & Fouladi-Nashta, A. A. (2016). "Localisation and endocrine control of hyaluronan synthase (HAS) 2, HAS3 and CD44 expression in sheep granulosa cells." *Reprod Fertil Dev*, 28 (6), 765-75. doi: 10.1071/RD14294.

Clarke, H. G., Hope, S. A., Byers, S. & Rodgers, R. J. (2006). "Formation of ovarian follicular fluid may be due to the osmotic potential of large glycosaminoglycans and proteoglycans." *Reproduction*, 132 (1), 119-31. doi: 10.1530/rep.1.00960.

Collins, S. L., Black, K. E., Chan-Li, Y., Ahn, Y. H., Cole, P. A., Powell, J. D. & Horton, M. R. (2011). "Hyaluronan fragments promote inflammation by down-regulating the anti-inflammatory A2a receptor." *Am J Respir Cell Mol Biol*, 45 (4), 675-83. doi: 10.1165/rcmb.2010-0387OC.

Conneely, O. (2000). "Progesterone receptors in reproduction: functional impact of the A and B isoforms." *Steroids*, 65 (10-11), 571-577. doi: 10.1016/s0039-128x(00)00115-x.

Cross, J. C., Werb, Z. & Fisher, S. J. (1994). "Implantation and the placenta: key pieces of the development puzzle." *Science*, 266 (5190), 1508-18. doi: 10.1126/science.7985020.

Danforth, D. N., Arthur Veis, Moira Breen, H. G. Weinstein, John C. Buckingham. & Pacita Manalo. (1974). "The effect of pregnancy and labor on the human cervix: Changes in collagen, glycoproteins, and glycosaminoglycans." *American Journal of Obstetrics and Gynecology*, 120 (5), 641-651. doi: 10.1016/0002-9378(74)90608-5.

Dattena, M., Mara, L., Aa Bin, T. & Cappai, P. (2007). "Lambing rate using vitrified blastocysts is improved by culture with BSA and hyaluronan." *Mol Reprod Dev*, 74 (1), 42-7. doi: 10.1002/mrd.20576.

DiGiulio, D. B., Callahan, B. J., McMurdie, P. J., Costello, E. K., Lyell, D. J., Robaczewska, A., Sun, C. L., Goltsman, D. S., Wong, R. J., Shaw, G., Stevenson, D. K., Holmes, S. P. & Relman, D. A. (2015). "Temporal and spatial variation of the human microbiota during pregnancy." *Proc Natl Acad Sci U S A*, 112 (35), 11060-5. doi: 10.1073/pnas.1502875112.

Dixon, C. L., Richardson, L., Sheller-Miller, S., Saade, G. & Menon, R. (2018). "A distinct mechanism of senescence activation in amnion epithelial cells by infection, inflammation, and oxidative stress." *Am J Reprod Immunol*, 79 (3). doi: 10.1111/aji.12790.

Danforth, D. N. (1947). "The fibrous nature of the human cervix and its relation to the isthmic segment in gravid and nongravid uteri. ." *Proc Inst Med Chic.*, 16 (10), 295.

Du, M. R., Zhou, W. H., Piao, H. L., Li, M. Q., Tang, C. L. & Li, D. J. (2012). "Cyclosporin A promotes crosstalk between human cytotrophoblast and decidual stromal cell through up-regulating CXCL12/CXCR4 interaction." *Hum Reprod*, 27 (7), 1955-65. doi: 10.1093/humrep/des111.

Dusio, G. F., Cardani, D., Zanobbio, L., Mantovani, M., Luchini, P., Battini, L., Galli, V., Diana, A., Balsari, A. & Rumio, C. (2011). "Stimulation of TLRs by LMW-HA induces self-defense mechanisms in vaginal epithelium." *Immunol Cell Biol*, 89 (5), 630-9. doi: 10.1038/icb.2010.140.

Fallacara, A., Baldini, E., Manfredini, S. & Vertuani, S. (2018). "Hyaluronic Acid in the Third Millennium." *Polymers (Basel)*, 10 (7). doi: 10.3390/polym10070701.

Ferrero, D. M., Larson, J., Jacobsson, B., Di Renzo, G. C., Norman, J. E., Martin, J. N., Jr. D'Alton, M., Castelazo, E., Howson, C. P., Sengpiel, V., Bottai, M., Mayo, J. A., Shaw, G. M., Verdenik, I., Tul,

N., Velebil, P., Cairns-Smith, S., Rushwan, H., Arulkumaran, S., Howse, J. L. & Simpson, J. L. (2016). "Cross-Country Individual Participant Analysis of 4.1 Million Singleton Births in 5 Countries with Very High Human Development Index Confirms Known Associations but Provides No Biologic Explanation for 2/3 of All Preterm Births." *PLoS One*, 11 (9), e0162506. doi: 10.1371/journal.pone.0162506.

Fouladi-Nashta, A. A., Raheem, K. A., Marei, W. F., Ghafari, F. & Hartshorne, G. M. (2017). "Regulation and roles of the hyaluronan system in mammalian reproduction." *Reproduction*, 153 (2), R43-R58. doi: 10.1530/REP-16-0240.

Furnus, C. C., de Matos, D. G. & Martinez, A. G. (1998). "Effect of hyaluronic acid on development of *in vitro* produced bovine embryos." *Theriogenology*, 49 (8), 1489-1499. doi: 10.1016/s0093-691x(98)00095-8.

Gardner, D. K., Rodriegez-Martinez, H. & Lane, M. (1999). "Fetal development after transfer is increased by replacing protein with the glycosaminoglycan hyaluronan for mouse embryo culture and transfer." *Hum Reprod*, 14 (10), 2575-80. doi: 10.1093/humrep/14.10.2575.

Gellersen, B., Fernandes, M. S. & Brosens, J. J. (2009). "Non-genomic progesterone actions in female reproduction." *Hum Reprod Update*, 15 (1), 119-38. doi: 10.1093/humupd/dmn044.

Godbole, G., Suman, P., Gupta, S. K. & Modi, D. (2011). "Decidualized endometrial stromal cell derived factors promote trophoblast invasion." *Fertil Steril*, 95 (4), 1278-83. doi: 10.1016/j.fertnstert.2010.09.045.

Golichowski, A. M., King, S. R. & Mascaro, K. (1980). "Pregnancy-related changes in rat cervical glycosaminoglycans." *Biochem J*, 192 (1), 1-8. doi: 10.1042/bj1920001.

Gonzalez, J. M., Xu, H., Ofori, E. & Elovitz, M. A. (2007). "Toll-like receptors in the uterus, cervix, and placenta: is pregnancy an immunosuppressed state?" *Am J Obstet Gynecol*, 197 (3), 296 e1-6. doi: 10.1016/j.ajog.2007.06.021.

Goshen, R., Ariel, I., Shuster, S., Hochberg, A., Vlodavsky, I., de Groot, N., Ben-Rafael, Z. & Stern, R. (1996). "Hyaluronan, CD44 and its variant exons in human trophoblast invasion and placental angiogenesis." *Mol Hum Reprod*, 2 (9), 685-91. doi: 10.1093/molehr/2.9.685.

Hadas, Ron, Eran Gershon, Aviad Cohen, Sima Stroganov, Ofir Atrakchi, Shlomi Lazar, Ofra Golani, Bareket Dassa, Michal Elbaz, Gadi Cohen, Elena Kartvelishvily, Raya Eilam, Nava Dekel. & Michal Neeman. (2020). "Hyaluronan-NK cell Interaction Controls the Primary Vascular Barrier during Early Pregnancy." doi: 10.1101/2020.02.09.940544.

Homandberg, G. A., Ummadi, V. & Kang, H. (2004). "The role of insulin-like growth factor-I in hyaluronan mediated repair of cultured cartilage explants." *Inflamm Res*, 53 (8), 396-404. doi: 10.1007/s00011-004-1276-y.

Irving-Rodgers, H. F. & Rodgers, R. J. (2005). "Extracellular matrix in ovarian follicular development and disease." *Cell Tissue Res*, 322 (1), 89-98. doi: 10.1007/s00441-005-0042-y.

Jang, G., Lee, B. C., Kang, S. K. & Hwang, W. S. (2003). "Effect of glycosaminoglycans on the preimplantation development of embryos derived from *in vitro* fertilization and somatic cell nuclear transfer." *Reprod Fertil Dev*, 15 (3), 179-85. doi: 10.1071/rd02054.

Jiang, D., Liang, J. & Noble, P. W. (2011). "Hyaluronan as an immune regulator in human diseases." *Physiol Rev*, 91 (1), 221-64. doi: 10.1152/physrev.00052.2009.

Kim, S. M. & Kim, J. S. (2017). "A Review of Mechanisms of Implantation." *Dev Reprod*, 21 (4), 351-359. doi: 10.12717/DR.2017.21.4.351.

Kimura, N., Konno, Y., Miyoshi, K., Matsumoto, H. & Sato, E. (2002). "Expression of hyaluronan synthases and CD44 messenger RNAs in porcine cumulus-oocyte complexes during *in vitro* maturation." *Biol Reprod*, 66 (3), 707-17. doi: 10.1095/biolreprod66.3.707.

Kolar, S. L., Kyme, P., Tseng, C. W., et al. (2015). "Group B Streptococcus Evades Host Immunity by Degrading Hyaluronan." *Cell Host Microbe.*, 18 (6), 694-704. doi: 10.1016/S0140-6736(13)60993-9.

Kultti, A., Pasonen-Seppanen, S., Jauhiainen, M., Rilla, K. J., Karna, R., Pyoria, E., Tammi, R. H. & Tammi, M. I. (2009). "4-Methylumbelliferone inhibits hyaluronan synthesis by depletion of cellular UDP-glucuronic acid and downregulation of hyaluronan synthase 2 and 3." *Exp Cell Res*, 315 (11), 1914-23. doi: 10.1016/j.yexcr.2009.03.002.

Kumar, D., Springel, E., Moore, R. M., Mercer, B. M., Philipson, E., Mansour, J. M., Mesiano, S., Schatz, F., Lockwood, C. J. & Moore, J. J. (2015). "Progesterone inhibits *in vitro* fetal membrane weakening." *Am J Obstet Gynecol*, 213 (4), 520 e1-9. doi: 10.1016/j.ajog.2015.06.014.

Lane, M., Maybach, J. M., Hooper, K., Hasler, J. F. & Gardner, D. K. (2003). "Cryo-survival and development of bovine blastocysts are enhanced by culture with recombinant albumin and hyaluronan." *Mol Reprod Dev*, 64 (1), 70-8. doi: 10.1002/mrd.10210.

Ledee, N., Munaut, C., Aubert, J., Serazin, V., Rahmati, M., Chaouat, G., Sandra, O. & Foidart, J. M. (2011). "Specific and extensive endometrial deregulation is present before conception in IVF/ICSI repeated implantation failures (IF) or recurrent miscarriages." *J Pathol*, 225 (4), 554-64. doi: 10.1002/path.2948.

Lockwood, C. J., Krikun, G., Hickey, M., Huang, S. J. & Schatz, F. (2009). "Decidualized human endometrial stromal cells mediate hemostasis, angiogenesis, and abnormal uterine bleeding." *Reprod Sci*, 16 (2), 162-70. doi: 10.1177/1933719108325758.

Lozovyy, V., Richardson, L., Saade, G. & Menon, R. (2021). "Progesterone receptor membrane components: key regulators of fetal membrane integrity." *Biol Reprod*, 104 (2), 445-456. doi: 10.1093/biolre/ioaa192.

MacDonald, P. C., Dombroski, R. A. & Casey, M. L. (1991). "Recurrent secretion of progesterone in large amounts: an endocrine/metabolic disorder unique to young women?" *Endocr Rev*, 12 (4), 372-401. doi: 10.1210/edrv-12-4-372.

Magier, S., van der Ven, H. H., Diedrich, K. & Krebs, D. (1990). "Significance of cumulus oophorus in *in-vitro* fertilization and oocyte viability and fertility." *Hum Reprod*, 5 (7), 847-52. doi: 10.1093/oxfordjournals.humrep.a137196.

Mahendroo, M. (2019). "Cervical hyaluronan biology in pregnancy, parturition and preterm birth." *Matrix Biol*, 78-79, 24-31. doi: 10.1016/j.matbio.2018.03.002.

Mansouri, M. R., Schuster, J., Badhai, J., Stattin, E. L., Losel, R., Wehling, M., Carlsson, B., Hovatta, O., Karlstrom, P. O., Golovleva,

I., Toniolo, D., Bione, S., Peluso, J. & Dahl, N. (2008). "Alterations in the expression, structure and function of progesterone receptor membrane component-1 (PGRMC1) in premature ovarian failure." *Hum Mol Genet*, 17 (23), 3776-83. doi: 10.1093/hmg/ddn274.

Marei, W. F., Salavati, M. & Fouladi-Nashta, A. A. (2013). "Critical role of hyaluronidase-2 during preimplantation embryo development." *Mol Hum Reprod*, 19 (9), 590-9. doi: 10.1093/molehr/gat032.

Marzioni, D., Crescimanno, C., Zaccheo, D., Coppari, R., Underhill, C. B. & Castellucci, M. (2001). "Hyaluronate and CD44 expression patterns in the human placenta throughout pregnancy." *Eur J Histochem*, 45 (2), 131-40. doi: 10.4081/1623.

McKeown-Longo, P. J. & Higgins, P. J. (2021). "Hyaluronan, Transforming Growth Factor beta, and Extra Domain A-Fibronectin: A Fibrotic Triad." *Adv Wound Care (New Rochelle)*, 10 (3), 137-152. doi: 10.1089/wound.2020.1192.

Menon, R., Boldogh, I., Hawkins, H. K., Woodson, M., Polettini, J., Syed, T. A., Fortunato, S. J., Saade, G. R., Papaconstantinou, J. & Taylor, R. N. (2014). "Histological evidence of oxidative stress and premature senescence in preterm premature rupture of the human fetal membranes recapitulated *in vitro*." *Am J Pathol*, 184 (6), 1740-51. doi: 10.1016/j.ajpath.2014.02.011.

Menon, R., Boldogh, I., Urrabaz-Garza, R., Polettini, J., Syed, T. A., Saade, G. R., Papaconstantinou, J. & Taylor, R. N. (2013). "Senescence of primary amniotic cells via oxidative DNA damage." *PLoS One*, 8 (12), e83416. doi: 10.1371/journal.pone.0083416.

Merlino, A., Welsh, T., Erdonmez, T., Madsen, G., Zakar, T., Smith, R., Mercer, B. & Mesiano, S. (2009). "Nuclear progesterone receptor expression in the human fetal membranes and decidua at term before and after labor." *Reprod Sci*, 16 (4), 357-63. doi: 10.1177/1933719108328616.

Mesiano, S. (2001). "Roles of estrogen and progesterone in human parturition." *Front Horm Res*, 27, 86-104. doi: 10.1159/000061038.

Mesiano, S., Chan, E. C., Fitter, J. T., Kwek, K., Yeo, G. & Smith, R. (2002). "Progesterone withdrawal and estrogen activation in human parturition are coordinated by progesterone receptor A expression in the myometrium." *J Clin Endocrinol Metab*, 87 (6), 2924-30. doi: 10.1210/jcem.87.6.8609.

Mesiano, S. & Welsh, T. N. (2007). "Steroid hormone control of myometrial contractility and parturition." *Semin Cell Dev Biol*, 18 (3), 321-31. doi: 10.1016/j.semcdb.2007.05.003.

Mitchell, B. F. & Taggart, M. J. (2009). "Are animal models relevant to key aspects of human parturition?" *Am J Physiol Regul Integr Comp Physiol*, 297 (3), R525-45. doi: 10.1152/ajpregu.00153.2009.

Mohey-Elsaeed, O., Marei, W. F. A., Fouladi-Nashta, A. A. & El-Saba, A. A. (2016). "Histochemical structure and immunolocalisation of the hyaluronan system in the dromedary oviduct." *Reprod Fertil Dev*, 28 (7), 936-947. doi: 10.1071/RD14187.

Kimura, N., Hoshino, Y., Totsukawa, K. & Sato, E. (2007). "Cellular and molecular events during oocyte maturation in mammals: molecules of cumulus-oocyte complex matrix and signalling pathways regulating meiotic progression." *Soc Reprod Fertil Suppl*, 63, 327-42.

Nadeem, L., Shynlova, O., Matysiak-Zablocki, E., Mesiano, S., Dong, X. & Lye, S. (2016). "Molecular evidence of functional progesterone withdrawal in human myometrium." *Nat Commun*, 7, 11565. doi: 10.1038/ncomms11565.

Nagyova, E. (2012). "Regulation of cumulus expansion and hyaluronan synthesis in porcine oocyte-cumulus complexes during *in vitro* maturation." *Endocr Regul*, 46 (4), 225-35. doi: 10.4149/endo_2012_04_225.

Nakagawa, K., Takahashi, C., Nishi, Y., Jyuen, H., Sugiyama, R., Kuribayashi, Y. & Sugiyama, R. (2012). "Hyaluronan-enriched transfer medium improves outcome in patients with multiple embryo transfer failures." *J Assist Reprod Genet*, 29 (7), 679-85. doi: 10.1007/s10815-012-9758-2.

Nakamura, K., Yokohama, S., Yoneda, M., Okamoto, S., Tamaki, Y., Ito, T., Okada, M., Aso, K. & Makino, I. (2004). "High, but not low, molecular weight hyaluronan prevents T-cell-mediated liver injury by reducing proinflammatory cytokines in mice." *J Gastroenterol*, 39 (4), 346-54. doi: 10.1007/s00535-003-1301-x.

Nallasamy, S. & Mahendroo, M. (2017). "Distinct Roles of Cervical Epithelia and Stroma in Pregnancy and Parturition." *Semin Reprod Med*, 35 (2), 190-200. doi: 10.1055/s-0037-1599091.

Ng, S. W., Norwitz, G. A., Pavlicev, M., Tilburgs, T., Simon, C. & Norwitz, E. R. (2020). "Endometrial Decidualization: The Primary Driver of Pregnancy Health." *Int J Mol Sci*, 21 (11). doi: 10.3390/ijms21114092.

Ohno-Nakahara, M., Honda, K., Tanimoto, K., Tanaka, N., Doi, T., Suzuki, A., Yoneno, K., Nakatani, Y., Ueki, M., Ohno, S., Knudson, W., Knudson, C. B. & Tanne, K. (2004). "Induction of CD44 and MMP expression by hyaluronidase treatment of articular chondrocytes." *J Biochem*, 135 (5), 567-75. doi: 10.1093/jb/mvh069.

Ohta, N., Saito, H., Kaneko, T., Yoshida, M., Takahashi, T., Saito, T., Nakahara, K. & Hiroi, M. (2001). "Soluble CD44 in human ovarian follicular fluid." *J Assist Reprod Genet*, 18 (1), 21-5. doi: 10.1023/a:1026494528415.

Olivares, E. G., Montes, M. J., Oliver, C., Galindo, J. A. & Ruiz, C. (1997). "Cultured human decidual stromal cells express B7-1 (CD80) and B7-2 (CD86) and stimulate allogeneic T cells." *Biol Reprod*, 57 (3), 609-15. doi: 10.1095/biolreprod57.3.609.

Patriarca, M. T., Barbosa de Moraes, A. R., Nader, H. B., Petri, V., Martins, J. R., Gomes, R. C. & Soares, J. M. Jr. (2013). "Hyaluronic acid concentration in postmenopausal facial skin after topical estradiol and genistein treatment: a double-blind, randomized clinical trial of efficacy." *Menopause*, 20 (3), 336-41. doi: 10.1097/GME.0b013e318269898c.

Peluso, J. J., Liu, X., Gawkowska, A. & Johnston-MacAnanny, E. (2009). "Progesterone activates a progesterone receptor membrane component 1-dependent mechanism that promotes human granulosa/luteal cell survival but not progesterone secretion." *J Clin Endocrinol Metab*, 94 (7), 2644-9. doi: 10.1210/jc.2009-0147.

Peluso, J. J., Liu, X., Gawkowska, A., Lodde, V. & Wu, C. A. (2010). "Progesterone inhibits apoptosis in part by PGRMC1-regulated gene expression." *Mol Cell Endocrinol*, 320 (1-2), 153-61. doi: 10.1016/j.mce.2010.02.005.

Peluso, J. J., Pappalardo, A., Losel, R. & Wehling, M. (2006). "Progesterone membrane receptor component 1 expression in the immature rat ovary and its role in mediating progesterone's antiapoptotic action." *Endocrinology*, 147 (6), 3133-40. doi: 10.1210/en.2006-0114.

Petrey, A. C. & de la Motte, C. A. (2014). "Hyaluronan, a crucial regulator of inflammation." *Front Immunol*, 5, 101. doi: 10.3389/fimmu.2014.00101.

Piao, H. L., Tao, Y., Zhu, R., Wang, S. C., Tang, C. L., Fu, Q., Du, M. R. & Li, D. J. (2012). "The CXCL12/CXCR4 axis is involved in the maintenance of Th2 bias at the maternal/fetal interface in early human pregnancy." *Cell Mol Immunol*, 9 (5), 423-30. doi: 10.1038/cmi.2012.23.

Rayahin, J. E., Buhrman, J. S., Zhang, Y., Koh, T. J. & Gemeinhart, R. A. (2015). "High and low molecular weight hyaluronic acid differentially influence macrophage activation." *ACS Biomater Sci Eng*, 1 (7), 481-493. doi: 10.1021/acsbiomaterials.5b00181.

Richardson, L. S., Taylor, R. N. & Menon, R. (2020). "Reversible EMT and MET mediate amnion remodeling during pregnancy and labor." *Sci Signal*, 13 (618). doi: 10.1126/scisignal.aay1486.

Rodgers, R. J. & Irving-Rodgers, H. F. (2010). "Formation of the ovarian follicular antrum and follicular fluid." *Biol Reprod*, 82 (6), 1021-9. doi: 10.1095/biolreprod.109.082941.

Rodgers, R. J., Irving-Rodgers, H. F. & Russell, D. L. (2003). "Extracellular matrix of the developing ovarian follicle." *Reproduction*, 126 (4), 415-24. doi: 10.1530/rep.0.1260415.

Rohe, H. J., Ahmed, I. S., Twist, K. E. & Craven, R. J. (2009). "PGRMC1 (progesterone receptor membrane component 1): a targetable protein with multiple functions in steroid signalling, P450 activation and drug binding." *Pharmacol Ther*, 121 (1), 14-9. doi: 10.1016/j.pharmthera.2008.09.006.

Romero, R., Hassan, S. S., Gajer, P., Tarca, A. L., Fadrosh, D. W., Nikita, L., Galuppi, M., Lamont, R. F., Chaemsaithong, P., Miranda, J., Chaiworapongsa, T. & Ravel, J. (2014). "The composition and stability of the vaginal microbiota of normal pregnant women is different from that of non-pregnant women." *Microbiome*, 2 (1), 4. doi: 10.1186/2049-2618-2-4.

Ruiz, C., Montes, M. J., Abadía-Molina, A. C. & Olivares, E. G. (1997). "Phagocytosis by fresh and cultured human decidual stromal cells:

opposite effects of interleukin-1 alpha and progesterone." *J Reprod Immunol*, 33 (1), 15-26.

Ruscheinsky, M., De la Motte, C. & Mahendroo, M. (2008). "Hyaluronan and its binding proteins during cervical ripening and parturition: dynamic changes in size, distribution and temporal sequence." *Matrix Biol*, 27 (5), 487-97. doi: 10.1016/j.matbio.2008.01.010.

Russell, D. L., Gilchrist, R. B., Brown, H. M. & Thompson, J. G. (2016). "Bidirectional communication between cumulus cells and the oocyte: Old hands and new players?" *Theriogenology*, 86 (1), 62-8. doi: 10.1016/j.theriogenology.2016.04.019.

Salustri, A., Yanagishita, M., Underhill, C. B., Laurent, T. C. & Hascall, V. C. (1992). "Localization and synthesis of hyaluronic acid in the cumulus cells and mural granulosa cells of the preovulatory follicle." *Dev Biol*, 151 (2), 541-51. doi: 10.1016/0012-1606(92)90192-j.

Sato, E. & Yokoo, M. (2005). "Morphological and biochemical dynamics of porcine cumulus-oocyte complexes: role of cumulus expansion in oocyte maturation." *Ital J Anat Embryol*, 110, 205-17.

Skarzynski, D. J., Piotrowska-Tomala, K. K., Lukasik, K., Galvao, A., Farberov, S., Zalman, Y. & Meidan, R. (2013). "Growth and regression in bovine corpora lutea: regulation by local survival and death pathways." *Reprod Domest Anim*, 48, Suppl 1, 25-37. doi: 10.1111/rda.12203.

Skinner, S. J. & Liggins, G. C. (1981). "Glycosaminoglycans and collagen in human amnion from pregnancies with and without premature rupture of the membranes.," 3 (2), 111-21. doi: *J Dev Physiol*.

Stock, A. E., Bouchard, N., Brown, K., Spicer, A. P., Underhill, C. B., Dore, M. & Sirois, J. (2002). "Induction of hyaluronan synthase 2 by human chorionic gonadotropin in mural granulosa cells of equine preovulatory follicles." *Endocrinology*, 143 (11), 4375-84. doi: 10.1210/en.2002-220563.

Straach, K. J., Shelton, J. M., Richardson, J. A., Hascall, V. C. & Mahendroo, M. S. (2005). "Regulation of hyaluronan expression during cervical ripening." *Glycobiology*, 15 (1), 55-65. doi: 10.1093/glycob/cwh137.

Tanghe, S., Van Soom, A., Nauwynck, H., Coryn, M. & de Kruif, A. (2002). "Minireview: Functions of the cumulus oophorus during oocyte maturation, ovulation, and fertilization." *Mol Reprod Dev*, 61 (3), 414-24. doi: 10.1002/mrd.10102.

Teklenburg, G., Salker, M., Molokhia, M., Lavery, S., Trew, G., Aojanepong, T., Mardon, H. J., Lokugamage, A. U., Rai, R., Landles, C., Roelen, B. A., Quenby, S., Kuijk, E. W., Kavelaars, A., Heijnen, C. J., Regan, L., Brosens, J. J. & Macklon, N. S. (2010). "Natural selection of human embryos: decidualizing endometrial stromal cells serve as sensors of embryo quality upon implantation." *PLoS One*, 5 (4), e10258. doi: 10.1371/journal.pone.0010258.

Thomas, P. (2008). "Characteristics of membrane progestin receptor alpha (mPRalpha) and progesterone membrane receptor component 1 (PGMRC1) and their roles in mediating rapid progestin actions." *Front Neuroendocrinol*, 29 (2), 292-312. doi: 10.1016/j.yfrne.2008.01.001.

Tienthai, P., Kimura, N., Heldin, P., Sato, E. & Rodriguez-Martinez, H. (2003). "Expression of hyaluronan synthase-3 in porcine oviducal epithelium during oestrus." *Reprod Fertil Dev*, 15 (1-2), 99-105. doi: 10.1071/rd02100.

Toyokawa, K., Harayama, H. & Miyake, M. (2005). "Exogenous hyaluronic acid enhances porcine parthenogenetic embryo development *in vitro* possibly mediated by CD44." *Theriogenology*, 64 (2), 378-92. doi: 10.1016/j.theriogenology.2004.12.005.

Ulbrich, S. E., Schoenfelder, M., Thoene, S. & Einspanier, R. (2004). "Hyaluronan in the bovine oviduct--modulation of synthases and receptors during the estrous cycle." *Mol Cell Endocrinol*, 214 (1-2):9-18. doi: 10.1016/j.mce.2003.12.002.

Urman, B., Yakin, K., Ata, B., Isiklar, A. & Balaban, B. (2008). "Effect of hyaluronan-enriched transfer medium on implantation and pregnancy rates after day 3 and day 5 embryo transfers: a prospective randomized study." *Fertil Steril*, 90 (3), 604-12. doi: 10.1016/j.fertnstert.2007.07.1294.

Vornhagen, J., Adams Waldorf, K. M. & Rajagopal, L. (2017). "Perinatal Group B Streptococcal Infections: Virulence Factors, Immunity, and

Prevention Strategies." *Trends Microbiol*, 25 (11), 919-931. doi: 10.1016/j.tim.2017.05.013.

Vornhagen, J., Quach, P., Boldenow, E., Merillat, S., Whidbey, C., Ngo, L. Y., Adams Waldorf, K. M. & Rajagopal, L. (2016). "Bacterial Hyaluronidase Promotes Ascending GBS Infection and Preterm Birth." *mBio*, 7 (3). doi: 10.1128/mBio.00781-16.

Wang, S., Sun, F., Han, M., Liu, Y., Zou, Q., Wang, F., Tao, Y., Li, D., Du, M., Li, H. & Zhu, R. (2019). "Trophoblast-derived hyaluronan promotes the regulatory phenotype of decidual macrophages." *Reproduction*, 157 (2), 189-198. doi: 10.1530/REP-18-0450.

Wu, W., Shi, S. Q., Huang, H. J., Balducci, J. & Garfield, R. E. (2011). "Changes in PGRMC1, a potential progesterone receptor, in human myometrium during pregnancy and labour at term and preterm." *Mol Hum Reprod*, 17 (4), 233-42. doi: 10.1093/molehr/gaq096.

Xu, B., Sun, X., Li, L., Wu, L., Zhang, A. & Feng, Y. (2012). "Pinopodes, leukemia inhibitory factor, integrin-beta3, and mucin-1 expression in the peri-implantation endometrium of women with unexplained recurrent pregnancy loss." *Fertil Steril*, 98 (2), 389-95. doi: 10.1016/j.fertnstert.2012.04.032.

Xu, H., Ito, T., Tawada, A., Maeda, H., Yamanokuchi, H., Isahara, K., Yoshida, K., Uchiyama, Y. & Asari, A. (2002). "Effect of hyaluronan oligosaccharides on the expression of heat shock protein 72." *J Biol Chem*, 277 (19), 17308-14. doi: 10.1074/jbc.M112371200.

Yokoo, M., Kimura, N. & Sato, E. (2010). "Induction of oocyte maturation by hyaluronan-CD44 interaction in pigs." *J Reprod Dev*, 56 (1), 15-9. doi: 10.1262/jrd.09-173e.

Yokoo, M. & Sato, E. (2011). "Physiological function of hyaluronan in mammalian oocyte maturation." *Reprod Med Biol*, 10 (4), 221-229. doi: 10.1007/s12522-011-0093-6.

Zhao, G., Zhou, X., Fang, T., Hou, Y. & Hu, Y. (2014). "Hyaluronic acid promotes the expression of progesterone receptor membrane component 1 via epigenetic silencing of miR-139-5p in human and rat granulosa cells." *Biol Reprod*, 91 (5), 116. doi: 10.1095/biolreprod.114.120295.

Zhu, R., Wang, S. C., Sun, C., Tao, Y., Piao, H. L., Wang, X. Q., Du, M. R. & Li Da-Jin. (2013). "Hyaluronan-CD44 interaction promotes growth of decidual stromal cells in human first-trimester pregnancy." *PLoS One*, 8 (9), e74812. doi: 10.1371/journal.pone.0074812.

In: Hyaluronic Acid
Editor: Vittorio Unfer
ISBN: 978-1-53619-743-3
© 2021 Nova Science Publishers, Inc.

Chapter 3

EFFECT OF HYALURONIC ACID TREATMENT IN MAINTAINING THE PHYSIOLOGICAL PREGNANCY: PRECLINICAL EVIDENCE

Serap Cilaker Micili[1,*], *PhD and Asli Goker*[2], *MD*
[1]Department of Histology and Embryology,
Dokuz Eylul University, Izmir, Turkey
[2]Department of Obstetrics and Gynecology,
Celal Bayar University, Manisa, Turkey

ABSTRACT

Hyaluronic acid (HA) is a nonsulfated glycosaminoglycan (GAG) found ubiquitously in the extracellular matrix of all mammalian tissues. HA has important roles in cell growth, adhesion, migration and differentiation. It has dynamic roles in the extracellular matrix of normal reproductive canal tissues, uterus during the menstrual cycle, reproductive system tissues and fluids during the fertilization process, blastocyst implantation process, and structure of the placenta and cervix during pregnancy and during delivery. This chapter will focus on the effects of hyaluronic

---

* Correspondence to: Serap Cilaker Micili, Department of Histology and Embryology, Dokuz Eylul University, Izmir, Turkey Email: cilaker6@yahoo.com.

acid treatment in maintaining the physiological pregnancy in preclinical studies.

Keywords: hyluronic acid, physiological pregnancy

1. INTRODUCTION

The extracellular matrix is composed of various biomolecules and specialized proteins such as laminin, fibronectin, elastin, collagen, and proteoglycans. Proteoglycans contain a core protein to which glycosaminglycan (GAG) chains are linked. Glycosaminoglycans are essential components of the extracellular matrix and contribute to cell recognition, adhesion, and growth regulation. They are classified according to whether they contain sulfates or not, these sulfates may be heparan sulfate, heparin, chondroitin sulfate, keratan sulfate I / II, dermatan sulfate and hyaluronic acid (HA) (Oliveria et al., 2015). Hyaluronic acid, also known as hyaluronan or hyaluronate, is a glycosaminoglycan that is naturally found throughout the body in many tissues and fluids and binds to specific cell surface receptors. It has distinct physicochemical properties of repeating β-1,4-D-glucuronic acid and β-1,3-N-acetylglucosamine units. Hyaluronic acid forms a polymer with this chemical structure. The size of this polymer ranges from 5,000 to 20,000,000 Daltons. Hyaluronic acid dimensions in the human body can vary according to the anatomical area where it is located. For example, it is 3-4 million Daltons in the synovial fluid in the joints and, 3,100,000 Daltons in the umbilical cord (Gupta et al., 2019). The higher concentration of HA is found in the connective tissue of tissues such as synovial fluid, skin, and vitreous humor while it is found in significant amounts in the lungs, brain, kidneys, and muscles, and it is rare in the liver. Its lower concentration is in the blood serum (Fallacara et al., 2018, Necas et al., 2008, Laurent et al., 1996).

Historically, HA was isolated for the first time as a glycosaminoglycan from the vitreous humor of the bovine eye. Meyer and Palmer named it "hyaluronic acid" (derived from hyaloid [vitreous] and uronic acid) in 1934. The term "hyaluronan" was introduced in 1986 to conform to polysaccharide nomenclature. The chemical structure of HA is C14H21NO11. After the first introduction in the vitreous humor, HA was isolated in many tissues in the body, it was found to be synthesized by all animals and in microbial (Streptococcus zooepidemicus, Escherichia coli, Bacillus subtilis, and others) fermentation, existing in all connective tissues of the body, including the skin, joint fluid, blood vessels, serum, brain, cartilage, heart valves, and umbilical cord (Liu et al., 2011).

The total volume of hyaluronic acid in the body is about 15 g for an adult weighing 70 kg, replacing one third of the hyaluronic acid after degradation and synthesis every day (Necas et al., 2008). Generally, it is produced by synoviocytes, fibroblasts, and chondrocytes. However, most cells in the body can synthesize HA at some point in their cell cycles, implicating its function in several fundamental biological processes (Falacara et al., 2018). HA has excellent viscoelasticity, high moisture retention capacity, high biocompatibility, and hygroscopic properties. At a concentration as low as 0.1%, HA chains can provide high viscosity (Laurent et al. 1996). By having these properties, HA acts as a lubricant, shock absorber, joint structure stabilizer, and water balance- and flow resistance-regulator (Conrozier et al., 2008, Ström et al., 2015). HA plays an important role in cell migration since it is involved in processes of growth, inflammation, and reparation as well as stimulation of different connective tissue cells. The physiological roles of HA are well-characterized in body tissues and fluids (Chernos et al., 2017).

In general, HA may be involved in various cellular interactions (cell differentiation, proliferation, development, and recognition) and physiological functions such as lubrication, hydration balance, matrix structure, and steric interactions. (George et al., 1998).

2. HYALURONIC ACID IN REPRODUCTIVE SYSTEM AND PREGNANCY

The involvement of GAG in female reproduction is of great importance and the role of GAGs in reproductive females continues to be investigated. Several animal studies have shown that GAGs are involved in physiological processes and are found in the uterus during the menstrual cycle and in the placenta during pregnancy (Oliveria et al., 2015). The main GAGs of the extracellular matrix of uterine tissues of mammals are chondroitin sulfate (CS), dermatan sulfate (DS), heparan sulfate (HS), and hyaluronic acid (HA) (Oliveira et al., 2015). Their role in pregnancy starts with the events associated with the fertilization process (Varner et al., 1991). HA is found in fluids and tissues of the reproductive system of humans, namely in the follicular fluid (Saito et al., 2000), amniotic fluid (Dahl et al., 1983), and serum (Kobayashi et al., 1999). Hyaluronic acid has been shown to play roles in the preparation of the endometrial stroma for embryonic implantation, differentiation of endometrial fibroblasts to decidual cells and cell proliferation (Martin et al., 2003). A study by Vandevoort shows that HA, namely high molecular weight hyaluronan, in the perivitelline space of oocytes (the space between zona pellucida and plasma membrane) contributes to the expansion of the perivitelline space by its water holding capacity (Vandevoort et al., 1997). Placental morphogenesis, functioning and regulation of the cardiovascular system including uteroplacental circulation and regulation of the systemic inflammatory response are the multidirectional effects of HA on pregnancy (Ziganshina et al., 2016). HA is also found in the stromal structures of the uterus and placenta and the angiogenic regions of decidua basalis (Goshen et al., 1996). Mesenchymal cell migration and formation of blood vessels are dependent on HA and villous stroma is stained positive for HA (Castelluci et al., 2000). Hyaluronan is shown to be present in cumulus cells, cervical mucus and oviductal, uterine and follicular fluid in humans (Suchanek et al. 1994). Cell surface transmembrane glycoprotein CD44 mediates the action of hyaluronan in human embryos and these receptors are expressed throughout the preimplantation period (Campbell et al. 1995).

Throughout pregnancy, the composition and concentration of GAGs change according to the needs of uterine and placental development, delivery, and postpartum involution of the uterus. Starting with the beginning of pregnancy, the uterus grows, and the cervix is rigid until the time of delivery. At the end of a term pregnancy, GAG levels are high and especially HA levels are increased (Golichowski et al., 1980, El Maradny et al., 1997). HA is a key molecule of the extracellular matrix, found whenever there is rapid tissue proliferation, regeneration, and repair. (Kobayashi et al. 1999). The extracellular matrix hyaluronic acid is produced in connective tissue fibroblasts located in the uterine cervix during parturition. The lower uterine segment and the amniotic fluid are also sources of hyaluronic acid (Kobayashi et al. 1999). At the moment of birth, HA concentrations are declined as a result of an increase in the activity of metalloproteinases which are enzymes that degrade components of the extracellular matrix and basal membrane. The change in HA concentration leads to leukocyte migration and dilatation of the cervix (El Maradny et al., 1997). Pathological opening of the cervix in case of infections is due to neutrophils causing a pro-inflammatory response (Holt et al., 2011). A study investigating the change in serum HA levels throughout pregnancy has shown that as the pregnancy reaches term, HA levels increase in maternal serum and this increase was found to be correlated to cervical ripening (Kobayashi et al. 1999). Throughout pregnancy, the cervix is dense and strong due to the stromal extracellular matrix and supports the fetüs. Ascending bacteria are prevented by a mucus barrier. As gestation reaches term, cervical ripening occurs which is remodeling of the extracellular matrix. An increase in HA synthesis, epithelial proliferation and changes in mucus are all necessary for a physiological delivery (Yellon et al., 2017, Read et al., 2007, Andersson et al. 2008, Mahendroo 1999).

Many species show an increased cervical hyaluronan synthesis in the cervical stroma, epithelium, and immune cells towards the end of pregnancy. Experiments in HA knock-out mice have shown that HA has a positive effect on fibrillary collagen modeling and arrangement in the lower genital tract. Similarly, HA has roles in establishing cellular permeability, epithelial tight connections and mucosal barriers to prevent infection. These roles have been shown in experiments in intestinal epithelial cells in which tight connection protein ZO-1 synthesis has been induced and so mice have been protected from infection (Kim et al, 2018). The connective tissue of the cervix is mainly composed of fibroblasts and the extracellular matrix contains collagen type I and III as well as elastin and fibronectin, water and proteoglycans. The cervical structure is constantly remodeled during pregnancy maintaining integrity with collagen fibers, proteoglycans, and hyaluronic acid (Junqueira et al. 1989).

Blastocyst implantation, preterm delivery, cervical ripening are some biological processes where HA is known to play essential roles. One of the most important reasons for preterm delivery which is a major determinant of neonatal mortality and morbidity is a bacterial infection (Ueno et al., 2015, Goldenberg et al., 2000, Nold et al., 2012). Bacteria that can pass through the cervical barrier and reach the amniotic cavity may increase pro-inflammatory cytokine levels, cause rupture of amniotic membranes, induce uterine contractions and facilitate cervical ripening which are all related to preterm delivery (Romero et al., 1992, Hilier et al., 1991). During this process remodeling of the extracellular matrix is an important step. High molecular weight HA polymers (HMW-HA) play an important role in cell migration and cell-to-cell signaling (Akgul et al., 2014). It is known that cervical HMW-HA is protective against preterm delivery due to its barrier function. An increase in the expression of hyaluronidase, an enzyme that breaks down HMW-HA, has been shown to increase preterm delivery in a murine infection model (Akgul et al., 2012, Mahendroo et al., 2012).

Not only bacterial but also viral infections may lead to preterm birth. The study by McGee et al., (2017) showed that HSV2 infection in a Mouse model resulted in cervical ripening with collagen remodeling and increased hyaluronic acid synthesis. The authors concluded that activation of Src kinase associated with HSV2 leads to increased HA synthesis in the epithelium which in turn affects collagen organization (McGee et al., 2017).

Cervical HA synthesis is evolutionarily conserved in mammals, showing an increase towards late pregnancy in humans, sheep, guinea pig, rabbit and rat (Nastha et al., 2012). Akgul et al. have measured this increase as from 19% of total GAG to 71% from early pregnancy to term, respectively (Akgul et al., 2012), and they found that the main HA synthase isoform responsible for cervical HA synthesis, was HAS2 (Akgul et al., 2014, Uchiyama et al., 2005). This phenomenon has also been shown in pregnant women who had upregulated HAS2 levels during labor (Straach et al., 2005). HAS2 is known to produce high molecular weight HA which increases the water content and cytokines of the cervix resulting in its ripening (El Maradny et al., 1997, Nastha et al., 2012). Cervical softening and increased distensibility at the beginning of labor are due to the increase in hyaluronic acid and water content and decrease in collagen. Cervical softening is the result of activation of matrix metalloproteinases (MMP-9 especially) that degrade collagen fibers (Facchinetti et al., 2005).

CONCLUSION

Hyaluronic acid, namely HMW-HA, is necessary throughout pregnancy and especially for maintaining cervical structure. Receptors such as hyaluronan binding proteins and hyaladherins are as important as HA itself. The endothelial glycocalyx is stabilised, vascular homeostasis is maintained and functions as a barrier (Ziganshina et al., 2016). All of these properties are important during pregnancy, but further studies are needed to provide insight into the function of HA and its receptors. Medications suitably for treatment with local HA may be the future goal for preventing preterm delivery.

REFERENCES

Akgul, Y., Word, R. A., Ensign, L. M., Yamaguchi Y., Lydon J., Hanes J., Mahendroo M. 2014. Hyaluronan in cervical epithelia protects against infection-mediated preterm birth. *J. Clin. Invest.* 124:5481–5489.

Akgul, Y., Holt, R., Mummert, M., Word, A., Mahendroo, M. 2012. Dynamic changes in cervical glycosaminoglycan composition during normal pregnancy and preterm birth. *Endocrinology* 153:3493–3503.

Andersson, S., Minjarez, D., Yost, N. P., Word, R. A. 2008. Estrogen and progesterone metabolism in the cervix during pregnancy and parturition. *The Journal of Clinical Endocrinology and Metabolism.* 93(6):2366–74.

Campbell, S., Swann, H. R., Aplin, J. D., et al., 1995. CD44 is expressed throughout preimplantation human embryo development. *Hum. Reprod.* 10:425– 430.

Castellucci M., Kosanke G., Verdenelli F., Huppertz B., Kaufmann P. 2000. Villous sprouting: fundamental mechanisms of human placental development. *Hum. Reprod. Update.* 6(5):485-94.

Chernos, M., Grecov, D., Kwok, E., Bebe S., Babsola O., Anastassiades T. 2017. Rheological study of hyaluronic acid derivatives. *Biomed. Eng. Lett.* 17;7(1):17-24.

Conrozier, T., Chevalier, X. 2008. Expert Opinion on Pharmacotherapy Long-term experience with hylan GF-20 in the treatment of knee osteoarthritis Drug Evaluation Long-term experience with hylan GF-20 in the treatment of knee osteoarthritis *Expert Opin. Pharmacother.*, 9(10):1797–1804.

Dahl, L. B., Hopwood, J. J., Laurent, U. B., Lilija, K., Tengblad, A. 1983 The concentration of hyaluronate in amniotic fluid. *Biochemical Medici*ne 30 280–283.

El Maradny, E., Kanayama, N., Kobayashi, H., Hossain, B., Khatun, S., Liping, S., et al., 1997. The role of hyaluronic acid as a mediator and regulator of cervical ripening. *Hum. Reprod.* 12(5): 1080-1088.

Facchinetti F., Venturini P., Blasi I., Giannella L. 2005 Changes in the cervical competence in preterm labour. *BJOG.* Mar;112 Suppl 1:23-7.

George E. 1998. Intra-articular hyaluronan treatment for osteoarthritis. *Ann. Rheum. Dis.* 57:637–40.

Goshen, R., Ariel, I., Shuster, S., Hochberg, A., Vlodavsky, I., de Groot, N., Ben-Rafael, Z., Stern, R. 1996. Hyaluronan, CD44 and its variant exons in human trophoblast invasion and placental angiogenesis. *Mol. Hum. Reprod.* 2(9):685-91.

Goldenberg, R. L., Hauth, J. C., Andrews, W. W. 2000. Intrauterine Infection and Preterm Delivery. *N. Engl. J. Med.* 342(20):1500–1507.

Golichowski, A. M., King, S. R., Mascaro, K. 1980. Pregnancy-related changes in rat cervical glycosaminoglycans. *Biochem.* 15;192(1):1-8.

Gupta, R. C., Lall, R., Srivastava, A., Sinha, A. 2019. Hyaluronic acid: Molecular mechanisms and therapeutic trajectory. *Front. Vet. Sci.* 25;6:192.

Hillier, S. L., Krohn, M. A., Kiviat, N. B., Watts, D. H., Eschenbach, D. A. 1991. Microbiologic causes and neonatal outcomes associated with chorioamnion infection. *Am. J. Obstet. Gynecol.* 165(4):955–961.

Holt, R., Timmons, B., Akgul, Y., Meredith, L., Akins, M. L., Mahendroo, M. 2011. The molecular mechanisms of cervical ripening differ between term and preterm birth. *Endocrinology.* 152(3):1036-46.

Junqueira, L. C., Zugaib, M., Montes, G. S., Toledo, O. M., Krisztan, R. M., Shigihara, K. M. 1989. Morphologic and biochemical evidence for the occurrence of collagenolysis and for the role of neutrophilic polymorphonuclear leukocytes during cervical dilatation. *Am. J. Obstet. Gynecol.* 138:273–281.

Kim, Y., West, G. A., Ray, Greeshma, et al., 2018. Layilin is critical for mediating hyaluronan 35 kDa-induced intestinal epithelial tight junction protein ZO-1 in vitro and in vivo. *Matrix Biol.* 66:93–109.

Kobayashi, H., Sun, G. W., Tanaka, Y., Terao, T. 1999. Serum hyaluronic acid levels during pregnancy and labor. *Obstet. Gynecol.* 93(4):480–484.

Laurent, T. C., Laurent, U. B. G., Fraser, J. R. E. 1996 The structure and function of hyaluronan: An overview. In: *Immunology and Cell Biology.* Nature Publishing Group; 74(2):a1–a7.

Liu, L., Liu, Y., Li J., Du, G., Chen, J. 2011. Microbial production of hyaluronic acid: current state, challenges, and perspectives *Microb. Cell Fact.* 10(1):99.

Mahendroo, M. 2012. Cervical remodeling in term and preterm birth: insights from an animal model *Reproduction* 143(4):429–438.

Mahendroo, M., Porter, A., Russell, D., Word, R. 1999. The Parturition Defect in Steroid 5α-Reductase Type 1 Knockout Mice Is Due to Impaired Cervical Ripening. *Molecular Endocrinology.*13:981–92.

McGee, D., Smith, A., Poncil S., Patterson, S., Bernstein, A. I., Racicot, K. 2017. Cervical HSV-2 infection causes cervical remodeling and increases risk for ascending infection and preterm birth. *PLoS ONE* 12(11): e0188645.

Martin, S. S., Soto-Suazo, M., Zorn, T. M. 2003. Distribution of versican and hyaluronan in the mouse uterus during decidualization *Bras. J. Med. Biol. Res.* 36(8)1067-71.

Nashta, A. A., Raheem, K. A., Marei, W. F., Ghafari, F., Hartshorne, G. M. 2017. Regulation and roles of the hyaluronan system in mammalian reproduction *Reproduction.* 153(2): R43-R58.

Necas, J., Bartosikova, L., Brauner, P., Kolar, J. 2008. Hyaluronic acid (hyaluronan): a review *Vet. Med. (Praha)*; 53(No. 8):397–411.

Nold, C., Anton, L., Brown, A., Elovitz, M. 2012. Inflammation promotes a cytokine response and disrupts the cervical epithelial barrier: a possible mechanism of premature cervical remodeling and preterm birth. *Am. J. Obstet. Gynecol.,* 206:208.e1–208e7.

Oliveira, G. B., Vale, A. B., Santoz, A. C. et al., 2015. Composition and significance of glycosaminoglycans in the uterus and placenta of mammals. *Brazilian Archives of Biology and Technology* 58(4):512-520.

Read, C. P., Word, R. A., Ruscheinsky, M. A., Timmons, B. C., Mahendroo, M. S. 2007. Cervical remodeling during pregnancy and parturition: molecular characterization of the softening phase in mice. *Reproduction.* 134(2):327–40.

Romero, R., Gonzalez, R., Sepulveda, W., Brandt, F., Ramirez, M., Sorokin, Y., Mazor, M., Treadwell, M. C., Cotton, D. B. 1992. Infection and labor. VIII. Microbial invasion of the amniotic cavity in patients with suspected cervical incompetence: prevalence and clinical significance. *Am. J. Obstet. Gynecol.* (12)80043 -3.

Saito, H., Kaneko T., Takahashi, T., Kawachiya, S., Saito, T., Hiroi, M. 2000. Hyaluronan in follicular fluids and fertilization of oocytes [Internet]. *Fertil. Steril.* 74(6):1148–1152.

Suchanek, E., Simunic, V., Juretic, D., Grizeli, V. 1994. Follicular fluid contents of hyaluronic acid, follicle-stimulating hormone and steroids relative to the success of in vitro fertilization of human oocytes. *Fertil. Steril.* 62: 347–352. 7.

Ström A., Larsson A., Okay O. 2015. Preparation and physical properties of hyaluronic acid-based cryogels. *J. Appl. Polym. Sci.* 132:42194.

Ueno, T., Niimi, H., Yonedo, N., Yoneda, S., et al., 2015. Eukaryote-made thermostable DNA polymerase enables rapid PCR-based detection of Mycoplasma, Ureaplasma and other bacteria in the amniotic fluid of preterm labor cases. *PLoS One* 4,;10(6):e0129032

VandeVoort, C. A., Cherr, G. N., Overstreet, J. W. 1997. Hyaluronic acid enhances the zona pellucida-induced acrosome reaction of macaque sperm. *J. Androl.* 18(1):1–5.

Varner, D. D., Forrest, D. W., Fuentes, F., Taylor, T. S., Hooper, R. N., Brinsko, S. P. et al., 1991. Measurements of glycosaminoglycans in follicular, oviductal and uterine fluids of mares. *J. Reprod. Fertil. Suppl.* 44: 297-306.

Straach, K. J., Shelton, J. M., Richardson, J. A., Hascall, V. C., Mahendroo. M. S. 2005 Regulation of hyaluronan expression during cervical ripening. *Glycobiology* 1555–65.

Uchiyama, T., Sakuta, T., Kanayama, T. 2005 Regulation of hyaluronan synthases in mouse uterine cervix. *Biochemical and Biophysical Research Communications* 327 927–932.

Yellon S. M. 2017. Contributions to the dynamics of cervix remodeling prior to term and preterm birth. *BOR.* 96(1):13–23.

Ziganshina, M. M., Pavlovich, S. S., Bovin, N. V., G. T. Sukhikh G. T. 2016. Hyaluronic Acid in Vascular and Immune Homeostasis during Normal Pregnancy and Preeclampsia. *Acta Naturae* 8;3:(30).

In: Hyaluronic Acid
Editor: Vittorio Unfer

ISBN: 978-1-53619-743-3
© 2021 Nova Science Publishers, Inc.

Chapter 4

SAFETY OF HYALURONIC ACID IN PREGNANCY

Giovanni Buzzaccarini[1,*], *Marco Noventa*[1]
and Antonio Simone Laganà[2]

[1]Department of Women's and Children's Health,
University of Padova, Padova, Italy
[2]Department of Obstetrics and Gynecology,
"Filippo del Ponte" Hospital,
University of Insubria, Varese, Italy

ABSTRACT

Hyaluronic acid (HA) is widely used for different purposes. Although biocompatible, a pregnancy risk-assessment has not been performed yet, but the high safety profile of HMW-HA further supports its administration in pregnancy. Indeed, several studies reported that HA administration is free of side effects. In this chapter, different routes of HA administration will be debated and an important focus will be given to HA safety for pregnancy. Conclusions suggest that well-designed randomized controlled trials are needed to clarify HA role in pregnancy but HA body

---

* Corresponding Author's Email: giovanni.buzzaccarini@gmail.com.

biocompatibility and its safety in previous studies should be the starting point for further trials.

INTRODUCTION

Hyaluronic acid (HA) is a biomolecule native for organisms. During the evolution of biological species, spanning millennia, its composition has undergone few modifications, suggesting a great and crucial role for life and evolution. For this reason, it seems reasonable to assume that it is unlikely to hamper cells, tissue or organs if administered in an exogenous way. However, there are some issues that need to be taken into consideration in order to assess HA safety. In particular, when giving concern to pregnancy, a high, and moreover, a certain safety profile is requested from national health systems and health insurances.

First of all, since HA is a medical device and not actually a chemical drug it may encounter lower accuracy in industrial manufacture and storage. Only few biopharmaceutical companies can provide a safe manufacturing profile, contrasting a widespread distribution of low-quality and impure products. These products can affect the organism in different ways, mainly related to components that can trigger an allergic reaction. However introduction of new technologies allows the production of highly pure HA without contaminants.

Secondly, HA can be administered as gel, tablets or an injectable solution. Regarding our clinical interest, the main ways of administration are topical, vaginal, intrauterine or injectable. Different ways of administration can lead to various locally HA concentrations and, subsequently, efficacy and/or hazard risk. Moreover, an injectable way is logically more traumatic than a topical application and can lead to bruising, hemorrhage, and infections.

Thirdly, due to the medical device classification, greater availability for non-medical healthcare professionals may occur. This can compromise safety and increase risks for patients who may unfortunately undergo illegal administration, especially when considering HA injectable fillers.

Fourthly, Paracelsus[1] in his *Defensio III*[2] stated *"Omnia venenum sunt: nec sine veneno quicquam existit. Dosis sola facit, ut venenum non sit*[3]*"* giving birth to the modern toxicology. The importance of this statement relies on the belief that everything can be harmful, depending on the dose. HA, although biocompatible, has not yet been proven to have a threshold dose, in fact as recently reviewed by Oe M. et al, in all the considered clinical trials no toxicity and no adverse events were recorded, even if HA was taken for periods up to 12 months at high concentration (Oe et al. 2016).

In this chapter we will discuss the main HA ways of administration, with a special focus on HA side effects and pregnancy safety. In particular, we consider not only ongoing pregnancy but also long-term outcomes such as fertility.

CORPUS

Vulvovaginal HA

Recently, our study group has been trying to perform a systematic review of HA local administration in vulva and vagina (Buzzaccarini 2021). Our main focus was specifically to identify all the studies which administered HA in vulva or vagina, independently from the condition or aim. We divided the studies into three groups:

1. Studies where HA was administered in postmenopausal women suffering from Vulvovaginal Atrophy (VVA), complaining of signs and symptoms related to it.
2. Studies where HA was administered in women who underwent radiotherapy for gynecological cancer (i.e., cervical cancer) and

---

[1] Paracelsus (1493/1494 – 24 September 1541; full name Philippus Aureolus Theophrastus Bombastus von Hohenheim) was a Swiss physician, alchemist, lay theologian, and philosopher of the Renaissance. He is believed to be the "father of toxicology".

[2] *Responsio ad quasdam accusationes et calumnias suorum aemulorum et obtrectatorum. Defensio III. Descriptionis et designationis nouorum Receptorum.*

[3] "Everything is poison: nothing exists that is non-poisonous. Only the dose prevents the poison from taking effect."

suffering from signs and symptoms related mainly to dyspareunia, bruising, and bleeding.
3. Studies where HA was administered in women of every age experiencing vulvar and vaginal signs and symptoms not properly related to VVA.

We identified 17 original studies administering HA: 8 randomized controlled studies (where controls were mainly estrogen therapies) and 9 longitudinal studies without controls. The majority of them used HA ovules, tablets or gel, and only few used injection techniques. However, very few studies administered HA alone and almost all of them used adjuvants. Our conclusion states that a well-designed randomized controlled study is needed to assess either HA efficacy and safety in vulvar and vaginal administration for VVA or other conditions.

Regarding HA side effects we found only few studies taking them into consideration. In particular, Condemi et al. (Condemi 2018) studied HA administration for vulvovaginal atrophy (VVA) patients. They performed a longitudinal non-controlled study administering HA with topical hyperbaric oxygen. One administration was performed 5 times every 2 weeks. Each session lasted 15 minutes with an oxygen flow of 2 L/min. Of 25 patients, only one woman experienced slight bleeding after the first sexual intercourse following the first treatment. This event was probably caused by new tissue vascularization.

Chen et al. (Chen 2013) performed a multicenter randomized controlled study. They enrolled 144 postmenopausal patients complaining of vaginal dryness. They divided them into two groups: in the first hyaluronic acid vaginal gel 5 g was applied once every 3 days for a total of 10 applications over 30 days; in the second, estriol cream 0,5 g was applied once every 3 days for a total of 10 applications over 30 days. A total of 13 adverse effects were reported in 144 participants, although only 4 were possibly related to HA administration: two women reported vulvovaginal candidiasis and one bacterial vaginitis which spontaneously resolved, and one vulvovaginal candidiasis which resolved with treatment.

Conversely, 8 studies analyzed but did not report adverse events (Serati 2015, Morali 2006, LeDonne 2011, Origoni 2016, Hersant 2018,

Quaranta 2014, Gonzalez 2019, Grimaldi 2012) suggesting the safety of HA administration.

As stated before, there are 7 other studies in literature, which administered HA in vulva or vagina. However, none of them investigated nor reported side effects (Ekin 2011, Delia 2019, Jokar 2016, Dinicola 2015, Costantino 2008, Palmieri 2019, Carter 2020).

Pregnancy was not an outcome considered for VVA postmenopausal patients. On the other hand, for patients undergoing radiotherapy for cervical cancer or complaining of vulvovaginal signs or symptoms, a pregnancy safety profile was not studied. In particular, a pregnant woman would benefit from HA administration if complaining of signs and symptoms as dyspareunia, bruising, pain, itching, bleeding, and urinary symptoms. However, since HA composition and biocompatibility has been extensively studied, and previous studies show no clearly identifiable side effects, it is therefore quite easy for further studies to use HA in the vulvovaginal district without great danger.

Intrauterine HA

The uterus has been considered a target for HA administration. Recent interest has been focused on intrauterine adhesions (IUAs), searching for possible therapeutic options to reduce their incidence. Indeed, IUAs, also known as Asherman syndrome, are fibrotic strings lining at opposite walls of the uterus and cervix. They can occupy the uterus cavity until a complete obliteration, impairing women's fertility (Dens 2010, Hooker 2014, Renier 2005). Moreover, they have quite various etiology, with a considerably stronger incidence after miscarriage treated by dilatation with blunt or suction curettage (Hooker 2014). If recorded, a hysteroscopy is needed to remove them. However, the recent scientific literature is interested in reducing its occurrence.

Quite recently, a multicenter prospective randomized controlled trial was performed (Hooker 2017). 152 women with a miscarriage of <14 weeks and at least one previous dilatation and/or curettage for miscarriage or termination of pregnancy. The intervention group (78) underwent an auto-crosslinked hyaluronic acid (ACP) gel administration, after dilatation and curettage.

This ACP gel was obtained by condensation of HA and it is reabsorbable in 7 days after intrauterine administration. This ACP gel (Hyalobarrier Gel Endo) was applied after dilatation and curettage with one single sterile syringe containing 10 mL of it and a 30-cm sterile cannula. The main result of this study showed that ACP gel administration reduced the incidence and severity of IUAs, decreased the mean adhesion scores, and decreased moderate and severe IUAs. However, since our main focus relies on HA safety, we must consider that this treatment is strictly linked to a post-natal application and, for this reason, not evaluable for pregnancy HA administration safety. Although, general complications from this procedure were similar for the intervention and control group and were mainly related to bleeding. Other complications reported were uterus perforation, cervix laceration, postoperative infection, pain and incomplete evacuation. However, none of these side effects were correlated to HA administration but mainly to surgical procedure.

More recently, a new multicenter, prospective, randomized trial tested another synthetic HA in preventing IUAs (Lee 2020). 192 subjects scheduled to undergo a hysteroscopic surgery were randomized into two groups. In the intervention group, 96 patients were administered a new synthetic HA called ABT13107; in the control group, Hyalobarrier Gel Endo was administered. A non-inferiority conclusion was drawn.

Adverse effects occurred in the 25.0% of cases for both groups but the adverse effects arising from the use of this device occurred in the 6.3% of the intervention group and the 3.1% of the control group. Adverse device effects in the intervention group were mainly due to an increase in low serum lipoprotein (LDL) concentration, followed by dysmenorrhea, pelvic pain, palpitation, abdominal pain, chest pain, and tendonitis. Similarly, in the control group, adverse device effects were mainly due to an increase in serum LDL concentration, pyrexia, and vaginal infection. However, all these adverse device effects were considered as either expected or not likely to be relevant to the HA device.

Similarly, one other study group (Li 2019) performed a randomized, double-blind, parallel-group, controlled multicenter trial where 300 women undergoing dilatation and curettage after a miscarriage were assigned to two groups. In the intervention group, 150 patients underwent a post-procedure application of hyaluronic acid gel (MateRegen®) composed of highly purified cross-linked hyaluronan chains. A significant reduction of IUAs was reported adding findings of intrauterine HA administration to a previous study. Blood loss was similar for both groups (intervention and control) and no signs of postoperative infection were reported in either group. Conversely, no side effects were attributed to HA administration.

A previous randomized controlled study (Tsapanos 2002) was conducted with similar procedures. A total of 150 patients were recruited with incomplete, missed, or recurrent abortion scheduled for dilatation and curettage. In the intervention group (50 patients) Seprafilm™ was administered after curettage. This device is a translucent bioresorbable membrane, biocompatible, nontoxic, and nonimmunogenic, composed of chemically modified hyaluronic acid (sodium hyaluronate) and carboxymethylcellulose (HA/CMC). It acts as a temporary adhesion barrier that keeps tissue surfaces separated during the early days of wound healing, when adhesions form. Seprafilm™ turns into a gel within 24–48 h after placement and then slowly resorbs 5–7 days. It is cleared from the body within 28 days. In this study, no adverse effect was reported. One other study (Diamond 1996) administered Seprafilm™ after laparoscopic myomectomy, showing a reduction in incidence, severity, extent, and area of postoperative uterine adhesions. No adverse effects, possibly related to the use of the membrane, were detected. More studies about Seprafilm™ showed no adverse effects related to the medical device. Indeed, the surgical procedures were the main cause of possible side effects (Beck 1997, Becker 1996). By contrast, a previous reported case of induced peritoneal inflammation was reported (Klingler 1999), although not related to a gynecological pathology.

Considering the long-term safety of intrauterine HA administration, we can have data from the study group of Hooker (Hooker 2020 1 and 2). In a 46-month follow-up after intrauterine application of ACP gel, significantly more ongoing pregnancies and live births were reported in women pursuing a pregnancy in the intervention group. Moreover, time to conception and time to conception resulting in live birth showed a shorter trend. Also, when corrected per age, the difference increased. On the other hand, the intervention group reported a reduction of dysmenorrhea and menstrual blood loss.

For previous studies, we can consider a systematic review and meta-analysis (Mais 2011) considering 5 randomized controlled trials administering Hyalobarrier gel. A total of 335 patients were included in all the studies. Regarding side effects, a total of 6 adverse events were reported, and all related to two trials that performed laparoscopy. In one of them, two patients had nausea and one patient had vomited in the intervention group and one patient had nausea in the control group. In the other trial, two patients in the control group had postoperative fever (<38.5°C). Conversely, no adverse events were reported in the three RCTs performed in hysteroscopy.

HA in IVF

In assisted reproductive techniques (ARTs), and especially in *In Vitro* Fertilization (IVF) and in Intra Cytoplasmic Sperm Injection (ICSI), HA has a limited but important role. In particular, during the Embryo Transfer (ET), the embryo is transferred into the uterus through a stick and inside a solution with different compounds. Traditionally, albumin was the main macromolecule of embryo culture media. Additionally, fibrin was inserted in the culture media for its adhesion properties. However, recently HMW-HA has been chosen for different properties:

1. HMW-HA is normally detected in endometrium and it has been proved in mice to dramatically increase on the day of embryo implantation (Carson 1987).
2. HMW-HA increases cell-to-cell adhesion and cell-to-matrix adhesion. This could prevent embryo expulsion and ectopic pregnancies.
3. HMW-HA is normally found in uterine, oviductal, and follicular fluids and normally produced by granulosa cells (Fancsovits 2015).

4. HMW-HA binds to CD44, which is normally expressed both on the embryo and in the endometrium stroma (Campbell 1995).
5. HMW-HA maintains viability on frozen embryos after thaw, increasing implantation rates after embryo transfer (Gardner 2003).

Adding together, these discoveries have greatly emphasized HMW-HA use in IVF. For this reason, new embryo media have been developed with higher HA concentration and lower albumin concentration. In particular, a new media has been produced with a fourfold increase in the concentration of hyaluronan (0.5 mg/mL vs. 0.125 mg/mL) and a fourfold decrease in the concentration of recombinant human albumin (2.5 mg/mL vs. 10 mg/mL) (EmbryoGlue, Vitrolife) (Urman 2008). Results from the first trials showed an increase in clinical pregnancy rates and implantation rates both for 3-day embryos and blastocysts. However, results were more prominent in women older than 35 years, in women with previous failed cycles of ART, and in women with poor-quality embryos (Urman 2008).

To analyze the real efficacy of HA, a Cochrane revision has been recently produced (Heymann 2020). 26 studies with a total of 6704 participants were enrolled in the review and meta-analysis. The certainty of evidence was primarily affected by imprecision and/or heterogeneity, determining a low to moderate quality. However, results were consistent showing that HMW-HA addition in the embryo transfer media showed an improved clinical pregnancy and live birth rates. These results were obtained considering moderate-quality evidence studies. However, when considering low-quality evidence studies it was also found a decrease in the miscarriage rates. Moreover, only 10 studies in this Cochrane review reported live birth rate, which is the most important clinical outcome. Probably, further studies should stress live birth rate as priority outcome. Summing together, evidence both from physiology and clinical trials supports HMW-HA usage in embryo culture media.

Cutaneous HA

Topical HA administration is one of the main routes of administration but, at the same time, one of the least efficient. Indeed, the great HA molecular weight makes it almost impossible for HA to enter the cutaneous layer (Jegasothy 2014). Hydration properties and wound healing are the main effects expected. A plethora of products are available in all chemists, all tested as topical cosmetical. In particular, Connettivina ® cream is prescribed for the treatment of skin disorders and dehydration. Moreover, it can be used in the prophylaxis of bacterial infections of traumatic lesions, skin sores and burns (Paghetti 2009). Taking into consideration our main interest, no studies have been performed assessing HA safety in pregnant women. However, since HA rarely enters the body through skin tissues and the hydration effect is limited to the surface, no adverse effect impairing pregnancy should be expected.

Cervical HA

Cervical HA administration is a new potential target therapy for preterm birth prevention. In fact, delivery before 37 weeks gestational age is one of the major causes of neonatal mortality and morbidity (Blencowe 2012 and 2013). Above all previous issues, cervical administration is the closest to pregnancy for HA studies. In fact, its target relies on cervical insufficiency, which is one significant cause of preterm birth. Actually, cervical insufficiency therapies provided are cerclage and tocolysis for reducing uterine contractions (Fonseca 2007). However, a greater interest has grown in cervical biology and, moreover, in possible therapies for increasing cervical volume. Starting from consideration of cervical ripening during labour, HA has been considered a possible target therapy. In fact, the cervix is principally composed of HA, collagen and proteoglycan. After the onset of labour, HA concentration increases markedly, and it is connected to an increase in water content. For this reason, hyaluronidase injections have been investigated as new method of cervical ripening induction and, subsequently, labour. Actually, it seems that hyaluronidase injections can be beneficial, but they are rarely used and, moreover, they could be refused by women (Kavanagh 2006). Conversely, cervical HA can be used with the opposite meaning: increasing water content and contrasting cervical ripening. For this reason, a cervical tissue model had been used to test this hypothesis. Unfortunately, cervical tissue from a pregnant woman is difficult to obtain, and cervical tissue samples from non-pregnant women were obtained.

To mimic cervical ripening, cervical tissue was enzymatically treated, which correlated to a decrease in collagen organization. This new softened tissue was then injected with crosslinked silk-HA composite hydrogel. The main results were a 54 ± 16% cervical volume increase and a boosted fibroblasts proliferation and metabolism for 5 days. Taken together, conclusions from this study provide an *ex vivo* pregnant-like tissue model for cervical ripening and suggest a need for well-designed trials with HA usage for increasing cervical bulking without stiffening (Raia 2020).

However, there is a strong need for *in vivo* trials in order to clarify the role of HA silk-hydrogel. Firstly, animal models are required cytotoxicity needs to be addressed. Secondly, pregnant women with cervical ripening should undergo a well-designed randomized controlled trial. In particular, side effects and preterm birth risk need to be carefully considered in order to maximize results.

CONCLUSION

Hyaluronic acid has been widely studied for different uses, related both to gynecological issues and not. There are few side effects and mainly related to the basal condition or the way of administration. Pregnancy is a condition where HA administration still finds few applications. For this reason, studies assessing HA safety profile are still lacking in the literature. Well-designed randomized controlled studies are needed for every HA purpose, assessing pregnancy safety both in acute and in long-term follow-up. Nevertheless, pieces of evidences are encouraging supporting HMW-HA administration in pregnancy and its safety profile.

REFERENCES

Acunzo, G; Guida, M; Pellicano, M; Tommaselli, GA; Di Spiezio Sardo, A; Bifulco, G; Cirillo, D; Taylor, A; Nappi, C. Effectiveness of auto-cross-linked hyaluronic acid gel in the prevention of intrauterine adhesions after hysteroscopic adhesiolysis: a prospective, randomized, controlled study. *Hum Reprod.*, 2003 Sep, 18(9), 1918-21. doi: 10.1093/humrep/deg368. PMID: 12923149.

Beck, ED. The role of Seprafilm™ bioresorbable membrane in adhesion prevention. *Eur J Surg*, 1997, Suppl 577, 49 –55.

Becker, JM; et al. Prevention of postoperative abdominal adhesions by a sodium hyaluronate-based bioresorbable membrane: A prospective, randomised, double blind multicenter study. *J Am Coll Surg*, 1996, 183, 297–306.

Blencowe, H; Cousens, S; Chou, D; Oestergaard, M; Say, L; Moller, AB; Kinney, M; Lawn, J. Born Too Soon Preterm Birth Action Group. Born too soon: the global epidemiology of 15 million preterm births. *Reprod Health.*, 2013, 10 Suppl 1, (Suppl 1), S2. doi: 10.1186/1742-4755-10-S1-S2. Epub 2013 Nov 15. PMID: 24625129, PMCID: PMC3828585.

Blencowe, H; Cousens, S; Oestergaard, MZ; Chou, D; Moller, AB; Narwal, R; Adler, A; Vera Garcia, C; Rohde, S; Say, L; Lawn, JE. National, regional, and worldwide estimates of preterm birth rates in the year 2010 with time trends since 1990 for selected countries: a systematic analysis and implications. *Lancet.*, 2012 Jun 9, 379(9832), 2162-72. doi: 10.1016/S0140-6736(12)60820-4. PMID: 22682464.

Buzzaccarini, G; Marin, L; Noventa, M; Vitagliano, A; Riva, A; Dessole, F; Capobianco, G; Bordin, L; Andrisani, A; Ambrosini, G. Hyaluronic acid in vulvar and vaginal administration: evidence from a literature systematic review. *Climacteric.*, 2021 Mar, 24, 1-12. doi: 10.1080/13697137.2021.1898580. Epub ahead of print. PMID: 33759670.

Campbell, S; Swann, HR; Aplin, JD; Seif, MW; Kimber, SJ; Elstein, M. CD44 is expressed throughout pre-implantation human embryo development. *Hum Reprod.*, 1995 Feb, 10(2), 425-30. doi: 10.1093/oxfordjournals.humrep.a135955. PMID: 7539449.

Carson, DD; Dutt, A; Tang, J. Glycoconjugate synthesis during early pregnancy: hyaluronate synthesis and function. *Developmental Biology*, 1987, 120, 228-35.

Carter J; Goldfarb, S; Baser, RE; Goldfrank, DJ; Seidel, B; Milli, L; Saban, S; Stabile, C; Canty, J; Gardner, GJ; Jewell, EL; Sonoda, Y; Kollmeier, MA; Alektiar, KM. A single-arm clinical trial investigating the effectiveness of a non-hormonal, hyaluronic acid-based vaginal moisturizer in endometrial cancer survivors. *Gynecol Oncol.*, 2020 Aug, 158(2), 366-374. doi: 10.1016/j.ygyno.2020.05.025. Epub 2020 Jun 8. PMID: 32522420, PMCID: PMC7423634.

Chen, J; Geng, L; Song, X; Li, H; Giordan, N; Liao, Q. Evaluation of the efficacy and safety of hyaluronic acid vaginal gel to ease vaginal dryness: a multicenter, randomized, controlled, open-label, parallel-group, clinical trial. *J Sex Med.*, 2013 Jun, 10(6), 1575-84. doi: 10.1111/jsm.12125. Epub 2013 Apr 9. PMID: 23574713.

Condemi, L; Di Giuseppe, J; Delli Carpini, G; Garoia, F; Frega, A; Ciavattini, A. Vaginal natural oxygenation device (VNOD) for concomitant administration of hyaluronic acid and topical hyperbaric oxygen to treat vulvo-vaginal atrophy: a pilot study. *Eur Rev Med Pharmacol Sci.*, 2018 Dec, 22(23), 8480-8486. doi: 10.26355/eurrev_201812_16548. PMID: 30556890.

Costantino, D; Guaraldi, C. Effectiveness and safety of vaginal suppositories for the treatment of the vaginal atrophy in postmenopausal women: an open, non-controlled clinical trial. *Eur Rev Med Pharmacol Sci.*, 2008 Nov-Dec, 12(6), 411-6. PMID: 19146203.

Deans, R; Abbott, J. Review of intrauterine adhesions. *J Minim Invasive Gynecol*, 2010, 17, 555–69.

Delia, P; Sansotta, G; Pontoriero, A; Iati, G; De Salvo, S; Pisana, M; Potami, A; Lopes, S; Messina, G; Pergolizzi, S. Clinical Evaluation of Low-Molecular-Weight Hyaluronic Acid-Based Treatment on Onset of Acute Side Effects in Women Receiving Adjuvant Radiotherapy after Cervical Surgery: A Randomized Clinical Trial. *Oncol Res Treat.*, 2019, 42(4), 217-223. doi: 10.1159/000496036. Epub 2019 Mar 12. PMID: 30861510.

Diamond, MD. Reduction of adhesions after uterine myomectomy by Seprafilm™ membrane (HAL-F): A blinded, prospective, randomised, multicenter clinical study. Seprafilm™ Adhesion Study Group. *Fertil Steril*, 1996, 66, 904 –910.

Dinicola, S; Pasta, V; Costantino, D; Guaraldi, C; Bizzarri, M. Hyaluronic acid and vitamins are effective in reducing vaginal atrophy in women receiving radiotherapy. *Minerva Ginecol.*, 2015 Dec, 67(6), 523-31. PMID: 26788875.

Ekin, M; Yaşar, L; Savan, K; Temur, M; Uhri, M; Gencer, I; Kıvanç, E. The comparison of hyaluronic acid vaginal tablets with estradiol vaginal tablets in the treatment of atrophic vaginitis: a randomized controlled trial. *Arch Gynecol Obstet.*, 2011 Mar, 283(3), 539-43. doi: 10.1007/s00404-010-1382-8. Epub 2010 Feb 5. PMID: 20135132.

Fancsovits, P; Lehner, A; Murber, A; Kaszas, Z; Rigo, J; Urbancsek, J. Effect of hyaluronan-enriched embryo transfer medium on IVF outcome: a prospective randomized clinical trial. *Arch Gynecol Obstet.*, 2015 May, 291(5), 1173-9. doi: 10.1007/s00404-014-3541-9. Epub 2014 Nov 15. PMID: 25398398.

Fonseca, EB; Celik, E; Parra, M; Singh, M; Nicolaides, KH. Fetal Medicine Foundation Second Trimester Screening Group. Progesterone and the risk of preterm birth among women with a short cervix. *N Engl J Med.*, 2007 Aug 2, 357(5), 462-9. doi: 10.1056/NEJMoa067815. PMID: 17671254.

Gardner, DK; Lane, M; Stevens, J; Schoolcraft, WB. Changing the start temperature and cooling rate in a slow-freezing protocol increases human blastocyst viability. *Fertil Steril.*, 2003 Feb, 79(2), 407-10. doi: 10.1016/s0015-0282(02)04576-4. PMID: 12568853.

González, IP; Leibaschoff, G; Esposito, C; Cipolla, G; Bader, A; Lotti, T; Tirant, M; Van Thuong, N; Ramo Abdulkader, M; Rauso, R; Serafin, D; Vojvodic, A; Zerbinati, N. Genitourinary syndrome of menopause and the role of biostimulation with non-cross-linked injectable hyaluronic acid plus calcium hydroxyapatite. *J Biol Regul Homeost Agents.*, 2019 Nov-Dec, 33(6), 1961-1966. doi: 10.23812/19-251-L. PMID: 31782291.

Grimaldi, EF; Restaino, S; Inglese, S; Foltran, L; Sorz, A; Di Lorenzo, G; Guaschino, S. Role of high molecular weight hyaluronic acid in postmenopausal vaginal discomfort. *Minerva Ginecol.*, 2012 Aug, 64(4), 321-9. PMID: 22728576.

Guida, M; Acunzo, G; Di Spiezio Sardo, A; Bifulco, G; Piccoli, R; Pellicano, M; Cerrota, G; Cirillo, D; Nappi, C. Effectiveness of auto-crosslinked hyaluronic acid gel in the prevention of intrauterine adhesions after hysteroscopic surgery: a prospective, randomized, controlled study. *Hum Reprod.*, 2004 Jun, 19(6), 1461-4. doi: 10.1093/humrep/deh238. Epub 2004 Apr 22. PMID: 15105384.

Hersant, B; SidAhmed-Mezi, M; Belkacemi, Y; Darmon, F; Bastuji-Garin, S; Werkoff, G; Bosc, R; Niddam, J; Hermeziu, O; La Padula, S; Meningaud, JP. Efficacy of injecting platelet concentrate combined with hyaluronic acid for the treatment of vulvovaginal atrophy in postmenopausal women with history of breast cancer: a phase 2 pilot study. *Menopause.*, 2018 Oct, 25(10), 1124-1130. doi: 10.1097/GME.0000000000001122. PMID: 29738415.

Heymann, D; Vidal, L; Or, Y; Shoham, Z. Hyaluronic acid in embryo transfer media for assisted reproductive technologies. *Cochrane Database Syst Rev.*, 2020 Sep 2, 9, CD007421. doi: 10.1002/14651858.CD007421.pub4. PMID: 32876946.

Hooker, AB; Lemmers, M; Thurkow, AL; Heymans, MW; Opmeer, BC; Br€olmann, HA; et al. Systematic review and meta-analysis of intrauterine adhesions after miscarriage: prevalence, risk factors and long-term reproductive outcome. *Hum Reprod Update*, 2014, 20, 262–78.

Hooker, AB; de Leeuw, R; van de Ven, PM; Bakkum, EA; Thurkow, AL; Vogel, NEA; van Vliet, HAAM; Bongers, MY; Emanuel, MH; Verdonkschot, AEM; Brölmann, HAM; Huirne, JAF. Prevalence of intrauterine adhesions after the application of hyaluronic acid gel after dilatation and curettage in women with at least one previous curettage: short-term outcomes of a multicenter, prospective randomized controlled trial. *Fertil Steril.*, 2017 May, 107(5), 1223-1231.e3. doi: 10.1016/j.fertnstert.2017.02.113. Epub 2017 Apr 6. PMID: 28390688.

Hooker, AB; de Leeuw, RA; Twisk, JWR; Brölmann, HAM; Huirne, JAF. Pregnancy and neonatal outcomes 42 months after application of hyaluronic acid gel following dilation and curettage for miscarriage in women who have experienced at least one previous curettage: follow-up of a randomized controlled trial. *Fertil Steril.*, 2020 Sep,

114(3), 601-609. doi: 10.1016/j.fertnstert.2020.04.021. Epub 2020 Jul 10. PMID: 32660725.

Hooker, AB; de Leeuw, RA; Twisk, JWR; Brölmann, HAM; Huirne, JAF. Reproductive performance of women with and without intrauterine adhesions following recurrent dilatation and curettage for miscarriage: long-term follow-up of a randomized controlled trial. *Hum Reprod.*, 2020 Dec 15, deaa289. doi: 10.1093/humrep/deaa289. Epub ahead of print. PMID: 33320197.

Jegasothy, SM; Zabolotniaia, V; Bielfeldt, S. Efficacy of a New Topical Nano-hyaluronic Acid in Humans. *J Clin Aesthet Dermatol.*, 2014 Mar, 7(3), 27-9. PMID: 24688623, PMCID: PMC3970829.

Jokar, A; Davari, T; Asadi, N; Ahmadi, F; Foruhari, S. Comparison of the Hyaluronic Acid Vaginal Cream and Conjugated Estrogen Used in Treatment of Vaginal Atrophy of Menopause Women: A Randomized Controlled Clinical Trial. *Int J Community Based Nurs Midwifery.*, 2016 Jan, 4(1), 69-78. PMID: 26793732, PMCID: PMC4709811.

Kavanagh, J; Kelly, AJ; Thomas, J. Hyaluronidase for cervical ripening and induction of labour. *Cochrane Database Syst Rev.*, 2006 Apr 19, (2), CD003097. doi: 10.1002/14651858.CD003097.pub2. PMID: 16625569.

Klingler, PJ; Floch, NR; Seelig, MH; Branton, SA; Wolfe, JT; Metzger, PP. Seprafilm™-induced peritoneal inflammation: A previously unknown complication. Report of a case. *Dis Colon Rectum*, 1999, 42, 1639 –1643.

Le Donne, M; Caruso, C; Mancuso, A; Costa, G; Iemmo, R; Pizzimenti, G; Cavallari, V. The effect of vaginally administered genistein in comparison with hyaluronic acid on atrophic epithelium in postmenopause. *Arch Gynecol Obstet.*, 2011 Jun, 283(6), 1319-23. doi: 10.1007/s00404-010-1545-7. Epub 2010 Jun 25. PMID: 20577750.

Lee, DY; Lee, SR; Kim, SK; Joo, JK; Lee, WS; Shin, JH; Cho, S; Park, JC; Kim, SH. A New Thermo-Responsive Hyaluronic Acid Sol-Gel to Prevent Intrauterine Adhesions after Hysteroscopic Surgery: A Randomized, Non-Inferiority Trial. *Yonsei Med J.*, 2020 Oct, 61(10), 868-874. doi: 10.3349/ymj.2020.61.10.868. PMID: 32975061, PMCID: PMC7515784.

Li, X; Wu, L; Zhou, Y; Fan, X; Huang, J; Wu, J; Yu, R; Lou, J; Yang, M; Yao, Z; Xue, M. New Crosslinked Hyaluronan Gel for the Prevention of Intrauterine Adhesions after Dilation and Curettage in Patients with Delayed Miscarriage: A Prospective, Multicenter, Randomized, Controlled Trial. *J Minim Invasive Gynecol.*, 2019 Jan, 26(1), 94-99. doi: 10.1016/j.jmig.2018.03.032. Epub 2018 Apr 17. PMID: 29678756.

Mais, V; Cironis, MG; Peiretti, M; Ferrucci, G; Cossu, E; Melis, GB. Efficacy of auto-crosslinked hyaluronan gel for adhesion prevention in laparoscopy and hysteroscopy: a systematic review and meta-analysis of randomized controlled trials. *Eur J Obstet Gynecol Reprod Biol.*, 2012 Jan, 160(1), 1-5. doi: 10.1016/j.ejogrb.2011.08.002. Epub 2011 Sep 25. PMID: 21945572.

Morali, G; Polatti, F; Metelitsa, EN; Mascarucci, P; Magnani, P; Marrè, GB. Open, non-controlled clinical studies to assess the efficacy and safety of a medical device in form of gel topically and intravaginally used in postmenopausal women with genital atrophy. *Arzneimittelforschung.*, 2006, 56(3), 230-8. doi: 10.1055/s-0031-1296715. PMID: 16618016.

Oe M, Tashiro T, Yoshida H, Nishiyama H, Masuda Y, Maruyama K, et al. Oral hyaluronan relieves knee pain: a review. *Nutr J* 2016 Jan 27;15:11-

Origoni, M; Cimmino, C; Carminati, G; Iachini, E; Stefani, C; Girardelli, S; Salvatore, S; Candiani, M. Postmenopausal vulvovaginal atrophy (VVA) is positively improved by topical hyaluronic acid application. A prospective, observational study. *Eur Rev Med Pharmacol Sci.*, 2016 Oct, 20(20), 4190-4195. PMID: 27831658.

Paghetti, A; Bellingeri, A; Pomponio, G; Sansoni, J; Paladino, D. Efficacia dell'acido ialuronico associato alla sulfadiazina argentica (Connettivina Plus) nel trattamento delle lesioni da pressione: uno studio di coorte osservazionale prospettico [Topic efficacy of ialuronic acid associated with argentic sulphadiazine (Connettivina Plus) in the treatment of pressure sores: a prospective observational cohort study]. *Prof Inferm.*, 2009 Apr-Jun, 62(2), 67-77. Italian. PMID: 19664355.

Palmieri, I. P. Biorevitalization of postmenopausal labia majora, the polynucleotide/hyaluronic acid option. *Obstet Gnecol Rep.*, 2019, 3 (2), doi: 10.15761/OGR.1000135.

Quaranta, L; Ottolina, J; Parma, M; Chionna, R; Sileo, F; Dindelli, M; Origoni, M; Candiani, M; Salvatore, S. An alternative approach for the treatment of vaginal atrophy. *Minerva Ginecol.*, 2014 Aug, 66(4), 377-81. PMID: 25020056.

Raia, NR; Bakaysa, SL; Ghezzi, CE; House, MD; Kaplan, DL. Ex vivo pregnant-like tissue model to assess injectable hydrogel for preterm birth prevention. *J Biomed Mater Res B Appl Biomater.*, 2020 Feb, 108(2), 468-474. doi: 10.1002/jbm.b.34403. Epub 2019 May 9. PMID: 31070848, PMCID: PMC7610206.

Renier D; Bellato P; Bellini D; Pavesio A; Pressato D; Borrione A. Pharmacokinetic behaviour of ACP gel, an autocrosslinked hyaluronan derivative, after intraperitoneal administration. *Biomaterials*, 2005, 26, 5368–74.

Serati, M; Bogani, G; Di Dedda, MC; Braghiroli, A; Uccella, S; Cromi, A; Ghezzi, F. A comparison between vaginal estrogen and vaginal hyaluronic for the treatment of dyspareunia in women using hormonal contraceptive. *Eur J Obstet Gynecol Reprod Biol.*, 2015 Aug, 191, 48-50. doi: 10.1016/j.ejogrb.2015.05.026. Epub 2015 Jun 3. PMID: 26070127.

Tsapanos, VS; Stathopoulou, LP; Papathanassopoulou, VS; Tzingounis, VA. The role of Seprafilm bioresorbable membrane in the prevention and therapy of endometrial synechiae. *J Biomed Mater Res.*, 2002, 63(1), 10-4. doi: 10.1002/jbm.10040. PMID: 11787023.

Urman, B., Yakin, K., Ata, B., Isiklar, A., Balaban, B. Effect of hyaluronan-enriched transfer medium on implantation and pregnancy rates after day 3 and day 5 embryo transfers: a prospective randomized study. *Fertil Steril.*, 2008 Sep, 90(3), 604-12. doi: 10.1016/j.fertnstert.2007.07.1294. Epub 2007 Oct 23. PMID: 17936283.

In: Hyaluronic Acid
Editor: Vittorio Unfer

ISBN: 978-1-53619-743-3
© 2021 Nova Science Publishers, Inc.

Chapter 5

USE OF HYALURONIC ACID IN ASSISTED REPRODUCTION TECHNIQUES

Maria Salomé Bezerra Espinola[1], Berniero Visconti[2] and Cesare Aragona[1,]*

[1]System Biology Group, Sapienza University of Rome, Italy
[2]University Tor Vergata, Rome, Italy

ABSTRACT

Since 1978, with the first successful *in vitro* fertilization (IVF), assisted reproductive technology (ART) has become an integral part of modern medicine, playing a pivotal role in achieving pregnancy and counteracting infertility. Despite these advanced techniques, the pregnancy and live birth success remain relatively low and it is estimated that implantation failure is the principal cause in 50%-70% of lost pregnancies. A growing body of evidence in the literature suggests a role for hyaluronic acid (hyaluronan, HA) in female reproduction. Hyaluronan can directly stimulate embryo growth or provide a more appropriate environment for the embryo to be nourished by supplements. In particular, embryo transfer using media with high concentrations of hyaluronic acid appears to increase the number of live births

---

* Corresponding Author's Email: aragonacesare@gmail.com.

compared to using solutions with low concentrations or without hyaluronic acid.

1. INTRODUCTION

Since the first successful *in vitro* fertilization (IVF) in 1978, assisted reproductive technology (ART) has become an integral part of modern medicine, playing a pivotal role in achieving pregnancy, and has revolutionized treatment options for couples with infertility (De Geyter 2019), a common condition worldwide with several medical, social and economic implications. The demand for ART has greatly increased, doubling its usage during the past decade as reported by Centers for Disease Control and Prevention (CDC). In general, there are 6 main components to an ART cycle: 1) Stimulation of the ovaries with exogenous gonadotropins (follicle-stimulating hormone (FSH) and luteinizing hormone (LH); 2) Inhibition of the natural spontaneous LH surge is required to minimize the likelihood of spontaneous ovulation prior to retrieval; 3) Oocyte retrieval using a transvaginal, ultrasound-guided approach; 4) Fertilization, embryo culture, and any micromanipulation procedures; 5) Embryo transfer, which involves the placement of embryos into the uterus through the cervix; 6) Luteal support most commonly consisting of exogenous progesterone, occasionally supplemented by estradiol (David A. Grainger 2013).

Despite these advanced techniques, pregnancy and live birth success remain relatively low and it is estimated that implantation failure is the principal cause in 50%-70% of lost pregnancies (De Geyter et al. 2018). Successful implantation depends on sequential and synchronized interactions between the embryo, while it is acquiring developmental competence, and the receptive uterus in a cyclical manner (Kliman and Frankfurter 2019). Understanding embryo metabolism has been a milestone over the past 30 years, in particular to develop more effective culture media, as well as enhanced media formulation to sustain prolonged embryo viability *in vitro*. Fertilization and embryo development *in vitro* introduce stresses which can impair not only embryo development but also its fate after the transfer (Lane and Gardner 1994). After all, *in vivo* developing embryo is exposed to several nutrients, hormones, cytokines and growth factors as it progresses to the uterus (Thouas et al. 2015) thus from the viewpoint of human ART, the major interest in chemical and physical factors lies in their adverse effects on the viability of embryos, even if exposure is for a relatively brief period of time (Wale PL 2016).

2. HYALURONAN IN ART

A growing body of evidence in the literature suggests a role for hyaluronic acid (hyaluronan, HA) in female reproduction while interest in its application in assisted reproduction is increasing (Heymann et al. 2020, Marei WFA 2017, Nakagawa et al. 2012, Urman et al. 2008, Sifer et al. 2009, Fancsovits et al. 2015, Loutradi et al. 2007, Nishihara and Morimoto 2017, Balaban B 2004). However, data on the efficacy of HA in assisted reproduction techniques and in particular on the effectiveness of its addition to the embryo transfer medium, to improve pregnancy rates, are still controversial (Safari et al. 2015).

HA is a member of the glycosaminoglycan family and a major component in uterine fluid. It has been shown that HA increases cell-cell and cell-matrix adhesion thus improving embryo apposition and attachment (Salustri 1999).

Implantation is a delicate process in which interaction among several factors from embryo or endometrium are involved and hence it is considered the major limiting step (Herrler, von Rango, and Beier 2003, Minas V 2005, Makker and Singh 2006b, Makrigiannakis A 2006). Recently, GAGs including HMW-HA were found in physiological uterine secretions (Zorn TM 1995, Yanagishita 1994;, Meinert et al. 2001) and this promptly induced researches to evaluate the ability of HMW-HA to increase the implantation rate of embryos (Gardner DK 1999).

It has been demonstrated that Hyaluronan significantly increases the day of implantation into the mouse uterus (Teixeira Gomes et al. 2009, Stojkovic et al. 2003) and appears to be associated with regions that contain proliferating stromal cells in preparation for embryo implantation (Carson DD 1987). Hyaluronan appears as a sticky viscous solution that resembles the natural secretions of the uterus. Studies on the distribution of HA in tissues are important to understand the histophysiological and pathological mechanisms underlying the events of the female reproductive tract, such as embryonic implantation, endometriosis, endometrial tumors. Moreover, Hyaluronic acid has been isolated in the uterine, oviductal and follicular fluids of mice, pigs, humans and cattle, and it has been also detected in oviductal fluids collected by catheterization during the oestrus cycle in heifers and cows, demonstrating that its concentration is higher during ovulation (Stojkovic et al. 2003, Fouladi-Nashta et al. 2017, Nagyova 2018, Bergqvist et al. 2005, Babayan et al. 2008, Gardner et al. 1996). Hyaluronic acid synthetase 2 (HAS2) and hyaluronic acid synthetase 3 (HAS3) transcripts have also been found in the oviducts of several animal species. HA synthesis is significantly increased in the uterus of mice on the day of implantation and the differential expression of HA in the human endometrium during the menstrual cycle suggests its involvement in implantation. In the human uterus, the peak expression of HAS and CD44 has been reported in the mid-secretory phase (Afify, Craig, and Paulino 2006) and evidence in the literature suggests the beneficial roles of HA in the implantation of human embryos (Schoolcraft W 2002, Balaban B 2004). Hyaluronan receptor has also been detected on the cell surface of human, bovine and porcine embryos and blastocysts (Kim et al. 2005). Histological preparations fixed and stained for hyaluronic acid and its CD44 receptor showed a peak during the secretory phase which coincides with the period in which the endometrium is most receptive to embryonic implantation.

2.1. Effect of Hyaluronan in the Medium of *In Vitro* Embryo Culture

Hyaluronan can directly stimulate embryo growth or provide a more appropriate environment for the embryo to be nourished by supplements (Gardner et al. 1996, Gardner 1998, Bavister 1995, Yanagishita 1994;, Zorn TM 1995). Supplementing the culture media with 0.5 mg/mL hyaluronic acid and 1.0 mg/mL heparin significantly increased the rate of blastocyst formation compared to the control group, suggesting a direct effect of hyaluronic acid on the *in vitro* growth of porcine embryos. Cell adhesion and cell-matrix adhesion are increased by hyaluronan as these mechanisms facilitate apposition and attachment of the blastocyst. Hyaluronan can also confer some protection to embryos during *in vitro* growth and cryopreservation. The addition of hyaluronan to culture medium containing bovine serum or recombinant albumin significantly increased the rate of blastocyst development and the ability of the blastocysts to survive after cryopreservation. In addition, the percentage of eight-cell embryos that re-expanded and hatched after freezing and thawing was also higher for embryos grown in the presence of hyaluronic acid, and birth rates in cows were also increased. Similar developmental abilities were observed in bovine embryos cultured in a synthetic oviduct culture medium containing both bovine serum albumin and hyaluronan (Schoolcraft W 2002, Balaban B 2004, Stojkovic et al. 2002, Chun et al. 2016).

Hyaluronan can also promote angiogenesis (West and Kumar 1989) both by its degradation products and by the interaction with the epidermal growth factor. Several studies have shown an increase in angiogenesis after hyaluronan administration. Hyaluronan and its CD44 receptor have been implicated in both angiogenesis and tumorigenesis in the human endometrium (Rooney et al. 1995). In the same way trophoblast invasion can be facilitated by hyaluronan.

The inhibition of adhesion of endometrial cells to the peritoneal mesothelium by hyaluronidase also suggests that the CD44-hyaluronan binding may be one of the mechanisms involved in the pathogenesis of endometriosis. On the other hand, the HA system is expressed in menstrual endometrial epithelial cells and endometrial stromal cells of women with and without endometriosis and the standard expression of CD44 does not differ in menstrual endometrial cells of women with and without endometriosis (Knudtson et al. 2019, Witz et al. 2003, Sarab Khlalaf Al-Juboory 2020).

3. PREPARATION OF THE UTERUS FOR IMPLANTATION: GENERAL AND RATIONALE CONCEPTS FOR THE USE OF HYALURONIC ACID IN ASSISTED REPRODUCTION TECHNIQUES

If the oocyte is fertilized, implantation occurs approximately eight days after ovulation, when the endometrial stroma is loose and edematous. In controlled ovarian hyperstimulation (COH) ART cycles, the luteal phase is usually induced and follows the administration of chorionic gonadotropin urinary or recombinant. The level of progesterone rapidly increases reaching maximum values within 3-5 days, when the cleaving embryo transfer (day 2-3) or blastocyst transfer (day 5-6) are performed. The rise of progesterone is also related to the amount of progesterone routinely administered to ART patients to counteract the impairment of the luteal phase following follicular aspiration. (Table 1)

Table 1. Serum progesterone and day of observation in 586 ART patients with transfer time ranged from 60 minutes after ICSI (Day 0) up to Day 5 (Blastocyst stage)

Time	N	Progesterone ng/ml (mean ± SD)	Implantation Rate %
at hCG	35	0.79 (±0.3)	20.1
at embryo transfer (Day 0)	35	8.5 (±2.3)	20.1
at embryo transfer (Day 1)	183	13 (±2.2)	20.3
at embryo transfer (Day 3)	278	27 (±3.2)	20.8
at embryo transfer (Day 5)	90	29 (±3.8)	20.5

(Aragona, personal observations)

Implantation considers three phases:

- Apposition: the blastocyst searches for the implant site. Here the so-called pinopods play an important role, helping the blastocyst to come into contact with the endometrial epithelium. They are morphological signals of endometrial receptivity to the

blastocyst, appearing only during the implantation window and disappearing around day 24 of the cycle.
- Adhesion: the blastocyst adheres to the endometrial epithelium and remains attached to it. This occurs between 6 and 7 days after fertilization, with the blastocyst already having a diameter of 300-400 µm.
- Invasion: the blastocyst (more specifically, the embryonic trophoblast) invades the endometrial stroma, breaks the basement membrane and penetrates the mother's blood vessels. Trophoblastic cells displace, dissociate and replace epithelial cells, eventually invading the basement membrane and the underlying stroma.

Implant failure is believed to be caused by inadequate receptivity in two thirds of cases and by problems of the embryo itself in the other third (Melford, Taylor, and Konje 2014). Recurrent implant failure is one of the most frequent causes of female infertility. Thus, pregnancy rates could be improved by optimizing endometrial receptivity. Evaluation of implantation markers can help to predict the outcome of assisted reproductive procedures and to detect any deficits early (Cakmak and Taylor 2011, Edgell, Rombauts, and Salamonsen 2013). Grosser at the beginning of 1910 stated that the most important physiological function of the endometrial cycle is the preparation for the reception of a fertilized egg (implant) (Grosser 1910). Croxatto et al. (Croxatto et al. 1978) found that, in the natural cycle, the human embryo usually arrives in the endometrial cavity 96 hours or more after the luteinizing hormone (LH) peak. He also reported the recovery of the ovum from the endometrial cavity 96 hours after the LH surge, and various authors have described a successful pregnancy after pronuclear embryo transfer the day after egg retrieval. In most IVF programs, fertilized embryos are transferred to the endometrial cavity 3-5 days after oocyte insemination (IVF) or intracytoplasmatic sperm injection (ICSI). In artificial cycles, as in egg donation or "freeze-all" programs, the transfer of embryos into the endometrial cavity between days 17 and 19 of the cycle has been reported to lead to a successful pregnancy (Li et al. 1991). The transfer of blastocysts into natural or replacement cycles in "freeze-all" programs seems to be the prevailing strategy used in routine clinical practice today. On the other hand, Aragona et al. reported full-term pregnancies and live births following the immediate uterine transfer of microinjected sperm oocytes in a natural cycle (Aragona et al. 2015a, Aragona et al. 2015b).

The microinjected oocyte, the oocyte in the pronuclear state and the embryo in the cleavage phase can all be accepted and nourished inside the uterine cavity where they may develop up to the blastocyst stage to be implanted when the implantation window becomes finally efficient. In this regard, the microinjected oocyte has been shown to be able to survive and spend more than 120 hours in the uterine cavity, before implantation, while the embryoblast spends approximately 72 hours in the uterine cavity before implantation. During this period, they cannot receive nutrition directly from the mother's blood and must rely on nutrients present in the uterine cavity, iron, fat-soluble vitamins, various steroid-dependent proteins, cholesterol and steroids) (Boron and Boulpaep). Implantation is further facilitated by the synthesis of matrix-associated substances, adhesion molecules and surface receptors for the matrix substance (Seppala 2004, Hempstock et al. 2004). (Table 2)

Table 2: Proteins, glycoproteins, peptides and factors found in the uterine cavity according to the literature:

Proteins, glycoproteins, peptides and factors found in the uterine cavity			
Matrix-associated:	**Others:**		
Fibronectin	Mucins	Beta-Lipoprotein	Beta-endorphin
Laminin	Prolactin	Relaxin	Leu-enkephalin
Entactin	IGFBP-1	Fibroblast growth factor 1	Diamine oxidase
Type IV-collagen	Placental protein 14(PP14) or glycodelin	Fibroblast growth factor 2	Tissue plasminogen activator
Heparan sulfate	Pregnancy-associated endometrial alpha-2-globulin (alpha-2-PEG)	Pregnancy-associated plasma protein A(PAPP-A)	Renin
Proteoglycan	Endometrial protein 15	Stress response protein 27(SRP-27)	Vit. D
Integrin	Albumin	CA-125	**Hyaluronic Acid**

Among these factors, endometrial hyaluronic acid received particular emphasis because it was significantly higher in women undergoing ovarian stimulation with letrozole plus FSH (158.7 ± 19.7 pg/ml) and Clomid plus FSH (160.8 ± 13.5 pg/ml) than in the control group (146.1 ± 26.2 pg/ml) (Sarab Khlalaf Al-Juboory 2020, Shadman 2009). A woman seemed to be able to influence the implantation potential of her embryos, in fact the average concentration of HA in follicular fluids was lower (158.0 ng/ml) in women whose embryos were not implanted, compared to the value found (220 ng/ml) in women who received one, 2 or 3 implanted embryos (239 ng/ml). Moreover, the level of HA was not correlated with maternal age, the number of oocytes collected or fertilized, or the number of embryos transferred. Follicular fluids from women with an endocrine problem had a significantly lower mean level of HA (142.0 ng/mL) than in women undergoing IVF due to male factor infertility (257.3 ng/mL) (Babayan et al. 2008).

4. HMW-HYALURONIC ACID IN EMBRYO TRANSFER MEDIUM (HETM) IN ASSISTED REPRODUCTION TECHNIQUES

Many embryonic culture media are commercially available for culturing pre-implantation human embryos in ART (Youssef et al. 2015). From the development of *in vitro* fertilization (IVF) techniques until today, embryonic culture media have evolved from salt solutions to a more complex media designed to address the metabolic and nutritional needs of specific phases of development and to mimic *in vivo* conditions (Urman et al. 2008, Bontekoe et al. 2010, Valojerdi et al. 2006).

Usually, the medium for embryo transfer is similar or identical to the solution for embryo growth, i.e., aqueous solutions enriched with 5-10% protein (serum albumin).

However, these media differ from the viscous fluid of the female reproductive tract so to simulate the chemical-physical properties of this fluid, and various macromolecules (for example, polyvinyl alcohol) have been added to the embryo culture media without increasing the implantation rate (I.R.).

The implantation is indeed a delicate process that involves complex interactions of various factors (Bavister 1995, Herrler, von Rango, and Beier 2003, Minas, Loutradis, and Makrigiannakis 2005, Makker and Singh 2006a, Makrigiannakis et al. 2006), and therefore remains the main limiting step.

In the first published study with hyaluronan-enriched transfer medium (HETM), Schoolcraft et al. reported significantly higher improved implantation rates (Schoolcraft W 2002). Furthermore, Gardner et al. demonstrated that HMW-HA can be added to embryonic culture media as an alternative to serum albumin (Gardner et al. 1996, Gardner 1998, Bavister 1995), improving implantation rates, cryotolerance and viability of the embryo after cryopreservation (Lane et al. 2003, Heymann et al. 2020, Hambiliki et al. 2010). Recently, GAGs including HMW-HA were found in physiological uterine secretions (Yanagishita 1994;, Zorn TM 1995, Meinert et al. 2001) prompting researchers to evaluate the ability of HMW-HA to increase the implantation rate of embryos. In the first published study with hyaluronan-enriched transfer medium (HETM), Schoolcraft et al. reported significantly higher improved implantation rates (Schoolcraft W 2002). Urman et al. obtained similar data, reporting improved implantation and clinical pregnancy rates for day 3 and day 5 transfers with HETM (Urman et al. 2008). They also reported an increased implantation rate with HETM in a subset of patients who had previously experienced implantation failures (Nakagawa et al. 2012). The effects of HETM on embryo growth in utero or embryo-endometrial crosstalk may be more important than its effects on the endometrium. Since implantation and multiple pregnancy rates have increased more markedly than clinical pregnancy rates with the use of HETM, it has been assumed that in the presence of a favorable endometrium, HETM increases the implantation potential of the transferred embryos. The increase in implantation rates observed in older women and in women with poor-quality embryos also supports this hypothesis. A woman's age is known to affect the quality of the oocytes and embryo, rather than the endometrium. The use of HETM in this group of women most likely increases implantation due to its effects on the embryo. The authors concluded that enrichment of the transfer medium with hyaluronic acid may benefit couples in assisted reproduction by transferring embryos at the cleavage or blastocyst stage. Overall implantation and clinical pregnancy rates were found to be significantly increased with HETM. Reducing the number of embryos transferred should be considered to avoid multiple pregnancies each time HETM is used. A beneficial effect was particularly evident in women> 35 years of age, women with premature ovarian failure and women receiving poor quality embryos.

From what has therefore been reported above in assisted reproduction procedures, it now appears that the embryo is transferred into the uterus preferably in a special transfer medium, a solution containing compounds that help the embryo to adhere successfully inside the uterus. Hyaluronic acid being a natural compound present in the body that acts as a binding and protective agent in tissues appears to play a role in mammalian reproduction (Fouladi-Nashta et al. 2017) and therefore it is often added to embryo transfer media to aid embryo implantation. The most adequate evaluation of the effectiveness of the use of HA in the embryo transfer medium was carried out in a study recently published in: "Cocrhane Database of Systematic Reviews 2020" (Heymann et al. 2020). In 26 studies involving 6704 women, aged between 27 to 35 years undergoing IVF/ICSI embryo transfer using media containing high concentrations of hyaluronic acid versus solutions containing no or low concentrations of HA, were compared in order to evaluate the possible effect of hyaluronan on the number of:

- Live births
- Spontaneous miscarriages (loss of pregnancy before the 20th week of gestation)
- Clinical pregnancies
- Multiple pregnancies
- Adverse (unwanted) events

With low/moderate evidence, embryo transfer using media with high concentrations of hyaluronic acid appears to increase the number of live births compared to using solutions with low concentrations or without hyaluronic acid (10 studies). If transfer media with low concentrations or without hyaluronic acid have a 33% chance of causing a live birth, solutions with high concentrations increase the chance of a live birth by up to 37% and 44%. The authors concluded that there would probably be 1 live birth additionally for every 14 embryos transferred in a highly concentrated hyaluronic acid solution. High concentrations of hyaluronic acid in the embryo transfer solution also appeared to increase the number of clinical pregnancies (17 studies) and the number of multiple pregnancies (7 studies). The use of transfer solutions containing high concentrations of hyaluronic acid appears to result in slightly fewer miscarriages (7 studies) No evidence was reported that the concentration of hyaluronic acid in the transfer solution affected the number of adverse events.

CONCLUSION

Based on the evidence in the literature, the effects induced by HETM are listed in Table 3.

Table 3. Effects of HETM

HETM, effects reported on:
1. Cleavage embryo/blastocyst
2. Endometrium
3. Crosstalk embryo - endometrium
4. Matrix cell adhesion and embryonic cells
5. Embryonic cryotolerance
6. Trophoblast invasion
7. Angiogenesis
8. Tumorogenesis endometrium

Although the exact biological mechanism underlying the role of hyaluronan on the implant is not fully understood, several hypotheses have been proposed to explain its beneficial effect. HA is one of the main components of the ECM, especially in tissues showing rapid growth and regeneration. Its production occurs at the uterine, oviductal and follicular level, with a sharp increase during pregnancy. Surface HA receptors have been found on human and bovine embryos, from oocyte to blastocyst, confirming the role of HA in embryonic implantation both as a structural molecule and as a regulatory factor. Several authors have described the ability of HMW-HA to protect blastocysts, increasing survival to cryopreservation when added to culture media. HMW-HA also increases cell-to-cell adhesion and cell-to-matrix adhesion, increasing blastocyst apposition and attachment. Trophoblast invasion may also be favored by HMW-HA, which facilitates cell migration and proliferation. Embryo transfer using solutions containing high concentrations of hyaluronic acid likely increases the number of live births in assisted reproductive techniques. Transfer solutions containing high concentrations of hyaluronic acid may slightly decrease the miscarriage rate. A positive effect on the birth rate remains to be defined. In fact, only ten of the 21 studies included in the Cochrane Database of Systematic Reviews reported this finding. The lack of studies reporting the number of live births was reported by the authors due to the large percentage of pregnancies that fail to progress to birth or may reflect the frequent practice of reporting studies before the last study participant has given birth. Other important observations not fully reported are related to spontaneous abortion, multiple pregnancies and other adverse events, such as ectopic pregnancy. Further controlled research on the actual functioning mechanism of the HA is in any case necessary as well as on the results related to the transfer of a single embryo.

REFERENCES

Afify, A. M., Craig, S. & Paulino, A. F. (2006). "Temporal variation in the distribution of hyaluronic acid, CD44s, and CD44v6 in the human endometrium across the menstrual cycle." *Appl Immunohistochem Mol Morphol*, 14 (3), 328-33. doi: 10.1097/00129039-200609000-00012.

Aragona, C., Linari, A., Micara, G., Tranquilli, D., Gambaro, A. M. & Bezerra Espinola, M. S. (2015a). "Term pregnancy and live birth subsequent to immediate uterine transfer of sperm microinjected oocyte in a natural cycle." *Gynecol Endocrinol*, 31 (8), 599-600. doi: 10.3109/09513590.2015.1031105.

Aragona, Cesare, Antonella Linari, Giulietta Micara, Daniela Tranquilli, Agnese Maria Lourdes Gambaro. & Maria Salome Bezerra Espinola. (2015b). "Successful use of "Hour 1" transfer of sperm microinjected oocytes in natural cycles of poor responder women: a prospective randomized controlled pilot study." *Gynecological Endocrinology*, 32 (5), 370-373. doi: 10.3109/09513590.2015.1121227.

Babayan, A., Neuer, A., Dieterle, S., Bongiovanni, A. M. & Witkin, S. S. (2008). "Hyaluronan in follicular fluid and embryo implantation following *in vitro* fertilization and embryo transfer." *J Assist Reprod Genet*, 25 (9-10), 473-6. doi: 10.1007/s10815-008-9268-4.

Balaban, B., Isiklar, A., Yakin, K., Gursoy, H. & Urman, B. (2004). "Increased implantation rates in patients with recurrent implantation failures following the use of a new transfer medium enriched with hyaluronan. ." *Hum Reprod*, 19, (i6–8).

Bavister, B. D. (1995). "Culture of preimplantation embryos: facts and artifacts." *Hum Reprod Update*, 1 (2), 91-148. doi: 10.1093/humupd/1.2.91.

Bergqvist, A. S., Yokoo, M., Heldin, P., Frendin, J., Sato, E. & Rodriguez-Martinez, H. (2005). "Hyaluronan and its binding proteins in the epithelium and intraluminal fluid of the bovine oviduct." *Zygote*, 13 (3), 207-18. doi: 10.1017/s0967199405003266.

Bontekoe, S., Blake, D., Heineman, M. J., Williams, E. C. & Johnson, N. (2010). "Adherence compounds in embryo transfer media for assisted reproductive technologies." *Cochrane Database Syst Rev*, 7 (7), CD007421. doi: 10.1002/14651858.CD007421.pub2.

Boron, Walter F. & Emile L Boulpaep. *Medical physiology: a cellular and molecular approach/[edited by] Walter F. Boron, Emile L. Boulpaep*. 2nd ed. ed: Philadelphia, PA: Saunders/Elsevier, c2009.

Cakmak, H. & Taylor, H. S. (2011). "Implantation failure: molecular mechanisms and clinical treatment." *Hum Reprod Update*, 17 (2), 242-53. doi: 10.1093/humupd/dmq037.

Carson, D. D., Dutt, A. & Tang, J. P. (1987). "Glycoconjugate synthesis during early pregnancy: hyaluronate synthesis and function." *Dev. Biol.*, 120, 228–35. doi: 10.1016/0012-1606(87)90120-5.

Chun, S., Seo, J. E., Rim, Y. J., Joo, J. H., Lee, Y. C. & Koo, Y. H. (2016). "Efficacy of hyaluronan-rich transfer medium on implantation and pregnancy rates in fresh and frozen-thawed blastocyst transfers in Korean women with previous implantation failure." *Obstet Gynecol Sci*, 59 (3), 201-7. doi: 10.5468/ogs.2016.59.3.201.

Croxatto, H. B., Ortiz, M. E., Díaz, S., Hess, R., Balmaceda, J. & Croxatto, H. D. (1978). "Studies on the duration of egg transport by the human oviduct." *American Journal of Obstetrics and Gynecology*, 132 (6), 629-634. doi: 10.1016/0002-9378(78)90854-2.

David A. Grainger, Bruce L. Tjaden. & Laura L. Tatpati. (2013). *Chapter 20 - Assisted Reproductive Technologies in Women and Health*. (Second Edition) ed.

De Geyter, C. (2019). "Assisted reproductive technology: Impact on society and need for surveillance." *Best Pract Res Clin Endocrinol Metab*, 33 (1), 3-8. doi: 10.1016/j.beem.2019.01.004.

De Geyter, C., Calhaz-Jorge, C., Kupka, M. S., Wyns, C., Mocanu, E., Motrenko, T., Scaravelli, G., Smeenk, J., Vidakovic, S., Goossens, V. & I. V. F. (2018). monitoring Consortium for the European Society of Human Reproduction European, and Embryology. "ART in Europe, 2014: results generated from European registries by ESHRE: The European IVF-monitoring Consortium (EIM) for the European Society of Human Reproduction and Embryology (ESHRE)." *Hum Reprod*, 33 (9), 1586-1601. doi: 10.1093/humrep/dey242.

Edgell, T. A., Rombauts, L. J. & Salamonsen, L. A. (2013). "Assessing receptivity in the endometrium: the need for a rapid, non-invasive test." *Reprod Biomed Online*, 27 (5), 486-96. doi: 10.1016/j.rbmo.2013.05.014.

Fancsovits, P., Lehner, A., Murber, A., Kaszas, Z., Rigo, J. & Urbancsek, J. (2015). "Effect of hyaluronan-enriched embryo transfer medium on IVF outcome: a prospective randomized clinical trial." *Arch Gynecol Obstet*, 291 (5), 1173-9. doi: 10.1007/s00404-014-3541-9.

Fouladi-Nashta, A. A., Raheem, K. A., Marei, W. F., Ghafari, F. & Hartshorne, G. M. (2017). "Regulation and roles of the hyaluronan system in mammalian reproduction." *Reproduction*, 153 (2), R43-R58. doi: 10.1530/REP-16-0240.

Gardner, David K. (1998). "Changes in requirements and utilization of nutrients during mammalian preimplantation embryo development and their significance in embryo culture." *Theriogenology*, 49 (1), 83-102. doi: 10.1016/s0093-691x(97)00404-4.

Gardner, David K., Michelle Lane, Ilan Calderon. & John Leeton. (1996). "Environment of the preimplantation human embryo *in vivo*: metabolite analysis of oviduct and uterine fluids and metabolism of cumulus cells**Supported by IVF America Inc., Greenwich, Connecticut and Monash IVF Pty. Ltd., Melbourne, Victoria, Australia." *Fertility and Sterility*, 65 (2), 349-353. doi: 10.1016/s0015-0282(16)58097-2.

Gardner, D. K., Rodriegez-Martinez, H. & Lane, M. (1999). "Fetal development after transfer is increased by replacing protein with the glycosaminoglycan hyaluronan for mouse embryo culture and transfer." *Hum Reprod.*, 14 ((10)), 2575-80. doi: 10.1093/humrep/14.10.2575. .

Grosser, O. (1910). The development of the egg membranes and the placenta; menstruation. Edited by P. Mall & W. Saunders. Eds J. Keibel, *In Manual of Human Embryology, Philadelphia and London in Book*

Hambiliki, F., Ljunger, E., Karlstrom, P. O. & Stavreus-Evers, A. (2010). "Hyaluronan-enriched transfer medium in cleavage-stage frozen-thawed embryo transfers increases implantation rate without improvement of delivery rate." *Fertil Steril*, 94 (5), 1669-73. doi: 10.1016/j.fertnstert.2009.10.019.

Hempstock, J., Cindrova-Davies, T., Jauniaux, E. & Burton, G. J. (2004). "Endometrial glands as a source of nutrients, growth factors and cytokines during the first trimester of human pregnancy: a morphological and immunohistochemical study." *Reprod Biol Endocrinol*, 2, 58. doi: 10.1186/1477-7827-2-58.

Herrler, Andreas, Ulrike von Rango. & Henning M. Beier. (2003). "Embryo-maternal signalling: how the embryo starts talking to its mother to accomplish implantation." *Reproductive BioMedicine Online*, 6 (2), 244-256. doi: 10.1016/s1472-6483(10)61717-8.

Heymann, D., Vidal, L., Or, Y. & Shoham, Z. (2020). "Hyaluronic acid in embryo transfer media for assisted reproductive technologies." *Cochrane Database Syst Rev*, 9, CD007421. doi: 10.1002/14651858.CD007421.pub4.

Kim, H. S., Lee, G. S., Hyun, S. H., Nam, D. H., Lee, S. H., Jeong, Y. W., Kim, S., Kim, J. H., Kang, S. K., Lee, B. C. & Hwang, W. S. (2005). "Embryotropic effect of glycosaminoglycans and receptors in development of porcine pre-implantation embryos." *Theriogenology*, 63 (4), 1167-80. doi: 10.1016/j.theriogenology.2004.06.001.

Kliman, H. J. & Frankfurter, D. (2019). "Clinical approach to recurrent implantation failure: evidence-based evaluation of the endometrium." *Fertil Steril*, 111 (4), 618-628. doi: 10.1016/j.fertnstert.2019.02.011.

Knudtson, J. F., McLaughlin, J. E., Santos, M. T., Binkley, P. A., Tekmal, R. R. & Schenken, R. S. (2019). "The Hyaluronic Acid System is Intact in Menstrual Endometrial Cells in Women With and Without Endometriosis." *Reprod Sci*, 26 (1), 109-113. doi: 10.1177/1933719118766257.

Lane, M. & Gardner, D. K. (1994). "Increase in postimplantation development of cultured mouse embryos by amino acids and induction of fetal retardation and exencephaly by ammonium ions." *J Reprod Fertil*, 102 (2), 305-12. doi: 10.1530/jrf.0.1020305.

Lane, M., Maybach, J. M., Hooper, K., Hasler, J. F. & Gardner, D. K. (2003). "Cryo-survival and development of bovine blastocysts are enhanced by culture with recombinant albumin and hyaluronan." *Mol Reprod Dev*, 64 (1), 70-8. doi: 10.1002/mrd.10210.

Li, T. C., Warren, M. A., Dockery, P. & Cooke, I. D. (1991). "Human endometrial morphology around the time of implantation in natural and artificial cycles." *J Reprod Fertil*, 92 (2), 543-54. doi: 10.1530/jrf.0.0920543.

Loutradi, K. E., Prassas, I., Bili, E., Sanopoulou, T., Bontis, I. & Tarlatzis, B. C. (2007). "Evaluation of a transfer medium containing high concentration of hyaluronan in human *in vitro* fertilization." *Fertil Steril*, 87 (1), 48-52. doi: 10.1016/j.fertnstert.2006.05.060.

Makker, A. & Singh, M. M. (2006a). "Endometrial receptivity: clinical assessment in relation to fertility, infertility, and antifertility." *Med Res Rev*, 26 (6), 699-746. doi: 10.1002/med.20061.

Makker, Annu. & Singh, M. M. (2006b). "Endometrial receptivity: Clinical assessment in relation to fertility, infertility, and antifertility." *Medicinal Research Reviews*, 26 (6), 699-746. doi: 10.1002/med. 20061.

Makrigiannakis, A., Minas, V., Kalantaridou, S. N., et al. (2006). "Hormonal and cytokine regulation of early implantation." *Trends Endocrinol Metab.*, 17 ((5)), 178-85. doi: 10.1016/j.tem.2006.05. 001..

Makrigiannakis, A., Minas, V., Kalantaridou, S. N., Nikas, G. & Chrousos, G. P. (2006). "Hormonal and cytokine regulation of early implantation." *Trends Endocrinol Metab*, 17 (5), 178-85. doi: 10. 1016/j.tem.2006.05.001.

Marei, W. F. A., Wathes, D. C., Raheem, K. A., Mohey-Elsaeed, O., Ghafari, F. & Fouladi-Nashta, A. A. (2017). "Influence of hyaluronan on endometrial receptivity and embryo attachment in sheep." *Reprod Fertil Dev.*, 29 ((9)), 1763-1773. doi: 10.1071/RD16232. PMID: 27725075.

Meinert, Mette, Gitte V. Eriksen, Astrid C. Petersen, Rikke B. Helmig, Claude Laurent, Niels Uldbjerg. & Anders Malmström. (2001). "Proteoglycans and hyaluronan in human fetal membranes." *American Journal of Obstetrics and Gynecology*, 184 (4), 679-685. doi: 10.1067/mob.2001.110294.

Melford, S. E., Taylor, A. H. & Konje, J. C. (2014). "Of mice and (wo)men: factors influencing successful implantation including endocannabinoids." *Hum Reprod Update*, 20 (3), 415-28. doi: 10. 1093/humupd/dmt060.

Minas, V., Loutradis, D. & Makrigiannakis, A. (2005). "Factors controlling blastocyst implantation." *Reprod Biomed Online*, 10 (2), 205-16. doi: 10.1016/s1472-6483(10)60942-x.

Minas, V., Loutradis, D. & Makrigiannakis, A. (2005). "Factors controlling blastocyst implantation." *Reprod Biomed Online*, 10 (2), 205-16. doi: 10.1016/s1472-6483(10)60942-x.

Nagyova, E. (2018). "The Biological Role of Hyaluronan-Rich Oocyte-Cumulus Extracellular Matrix in Female Reproduction." *Int J Mol Sci*, 19 (1). doi: 10.3390/ijms19010283.

Nakagawa, K., Takahashi, C., Nishi, Y., Jyuen, H., Sugiyama, R., Kuribayashi, Y. & Sugiyama, R. (2012). "Hyaluronan-enriched transfer medium improves outcome in patients with multiple embryo transfer failures." *J Assist Reprod Genet*, 29 (7), 679-85. doi: 10.1007/s10815-012-9758-2.

Nishihara, T. & Morimoto, Y. (2017). "Evaluation of transfer media containing different concentrations of hyaluronan for human *in vitro* fertilization." *Reprod Med Biol*, 16 (4), 349-353. doi: 10.1002/rmb2.12051.

Rooney, P., Kumar, S., Ponting, J. & Wang, M. (1995). "The role of hyaluronan in tumour neovascularization (review)." *Int J Cancer*, 60 (5), 632-6. doi: 10.1002/ijc.2910600511.

Safari, S., Razi, M. H., Safari, S. & Razi, Y. (2015). "Routine use of EmbryoGlue((R)) as embryo transfer medium does not improve the ART outcomes." *Arch Gynecol Obstet*, 291 (2), 433-7. doi: 10.1007/s00404-014-3416-0.

Salustri, A. (1999). "Hyaluronan and proteoglycans in ovarian follicles." *Human Reproduction Update*, 5 (4), 293-301. doi: 10.1093/humupd/5.4.293.

Sarab Khlalaf Al-Juboory, Mufeda Ali Jwad. & Muayad Sraibet Abood. (2020). "Relation of Endometrial Hyaluronic Acid with Female Infertility in Women Undergoing Ovarian Stimulation Protocol" *Indian Journal of Forensic Medicine & Toxicology*, 14 (3). doi: https://doi.org/10.37506/ijfmt.v14i3.10832.

Schoolcraft, W., Lane, M., Stevens, J. & Gardner, D. (2002). "Increased hyaluronan concentration in the embryo transfer medium results in a significant increase in human embryo implantation rate." *Fertil Steril*, 78 (S5.).

Seppala, M. (2004). "Advances in uterine protein research: reproduction and cancer." *Int J Gynaecol Obstet*, 85 (2), 105-18. doi: 10.1016/j.ijgo.2004.01.007.

Shadman, Sedigheh Dehbashi Sara Dehbashi Talieh Kazerooni Minoo Robati Saeed Alborzi Mohammad Ebrahim Parsanezhad Arash. (2009). "*Comparison of the Effects of Letrozole and Clomiphene*

Citrate on Ovulation and Pregnancy Rate in Patients with Polycystic Ovary Syndrome.," 34 (1), 23-28.

Sifer, C., Mour, P., Tranchant, S., Visentin, E., Hafhouf, E., Sermondade, N., Martin-Pont, B., Benzacken, B. & Levy, R. (2009). "[Is there an interest in the addition of hyaluronan to human embryo culture in IVF/ICSI attempts?]." *Gynecol Obstet Fertil*, 37 (11-12), 884-9. doi: 10.1016/j.gyobfe.2009.09.017.

Stojkovic, M., Kolle, S., Peinl, S., Stojkovic, P., Zakhartchenko, V., Thompson, J. G., Wenigerkind, H., Reichenbach, H. D., Sinowatz, F. & Wolf, E. (2002). "Effects of high concentrations of hyaluronan in culture medium on development and survival rates of fresh and frozen-thawed bovine embryos produced *in vitro*." *Reproduction*, 124 (1), 141-53.

Stojkovic, M., Krebs, O., Kolle, S., Prelle, K., Assmann, V., Zakhartchenko, V., Sinowatz, F. & Wolf, E. (2003). "Developmental regulation of hyaluronan-binding protein (RHAMM/IHABP) expression in early bovine embryos." *Biol Reprod*, 68 (1), 60-6. doi: 10.1095/biolreprod.102.007716.

Teixeira Gomes, R. C., Verna, C., Nader, H. B., dos Santos Simoes, R., Dreyfuss, J. L., Martins, J. R., Baracat, E. C., de Jesus Simoes, M. & Soares, J. M. Jr. (2009). "Concentration and distribution of hyaluronic acid in mouse uterus throughout the estrous cycle." *Fertil Steril*, 92 (2), 785-92. doi: 10.1016/j.fertnstert.2008.07.005.

Thouas, G. A., Dominguez, F., Green, M. P., Vilella, F., Simon, C. & Gardner, D. K. (2015). "Soluble ligands and their receptors in human embryo development and implantation." *Endocr Rev*, 36 (1), 92-130. doi: 10.1210/er.2014-1046.

Urman, B., Yakin, K., Ata, B., Isiklar, A. & Balaban, B. (2008). "Effect of hyaluronan-enriched transfer medium on implantation and pregnancy rates after day 3 and day 5 embryo transfers: a prospective randomized study." *Fertil Steril*, 90 (3), 604-12. doi: 10.1016/j.fertnstert.2007.07.1294.

Valojerdi, M. R., Karimian, L., Yazdi, P. E., Gilani, M. A., Madani, T. & Baghestani, A. R. (2006). "Efficacy of a human embryo transfer medium: a prospective, randomized clinical trial study." *J Assist Reprod Genet*, 23 (5), 207-12. doi: 10.1007/s10815-006-9031-7.

Wale, P. L. & Gardner, D. K. (2016). "The effects of chemical and physical factors on mammalian embryo culture and their importance for the practice of assisted human reproduction." *Hum Reprod Update*, 22 (1), 2-22. . doi: 10.1093/humupd/dmv03.

West, D. C. & Kumar, S. (1989). "Hyaluronan and angiogenesis." *Ciba Found Symp*, 143, 187-201, discussion 201-7, 281-5. doi: 10.1002/9780470513774.ch12.

Witz, C. A., Allsup, K. T., Montoya-Rodriguez, I. A., Vaughan, S. L., Centonze, V. E. & Schenken, R. S. (2003). "Pathogenesis of endometriosis--current research." *Hum Fertil (Camb)*, 6 (1), 34-40. doi: 10.1080/1464770312331368973.

Yanagishita, M. (1994). "Proteoglycans and hyaluronan in female reproductive organs." *EXS.*, 70, 179-90. doi: 10.1007/978-3-0348-7545-5_10. .

Youssef, Mohamed M. A., Eleni Mantikou, Madelon van Wely, Fulco Van der Veen, Hesham G. Al-Inany, Sjoerd Repping. & Sebastiaan Mastenbroek. (2015). "Culture media for human pre-implantation embryos in assisted reproductive technology cycles." *Cochrane Database of Systematic Reviews.* doi: 10.1002/14651858.CD007876.pub2.

Zorn, T. M., Pinhal, M. A., Nader, H. B., Carvalho, J. J., Abrahamsohn, P. A. & Dietrich, C. P. (1995). "Biosynthesis of glycosaminoglycans in the endometrium during the initial stages of pregnancy of the mouse." *Cell Mol Biol (Noisy-le-grand).*, 41 (1), 97-106.

In: Hyaluronic Acid
Editor: Vittorio Unfer
ISBN: 978-1-53619-743-3
© 2021 Nova Science Publishers, Inc.

Chapter 6

ROLE OF HYALURONAN IN FETAL DEVELOPMENT

Cora M. Demler and Natasza A. Kurpios[*]
Department of Molecular Medicine, College of Veterinary Medicine,
Cornell University, Ithaca, NY 14853, US

ABSTRACT

Cells sense their environment and respond to it, modulating the ECM by secreting structural, signaling, and enzymatic ECM components. A developing embryo is incredibly dynamic, and thus the ECM of its tissues must also change to accommodate proper morphogenesis. A key component of these soft ECM is hyaluronic acid (HA), which is essential for development from the fertilization of the egg until after birth. The chapter will illustrate the critical role that HA synthases, hyaluronidases, receptors, hyaladherins, and modifiers have on a variety of developmental processes, from early events like embryo implantation to post-natal acclimation to breathing and feeding. HA may appear to be a simple sugar, but its effects on vertebrate life are fascinatingly complex and it has critical roles in embryonic development in which a balanced regulation at the level of its synthesis, degradation, receptor

---

[*] Corresponding Author's Email: natasza.kurpios@cornell.edu.

binding, and proteoglycan binding, can regulate cardiac, intestinal morphogenesis, vasculogenesis and neonatal development.

Keywords: hyaluronic acid, developing embryo, cardiac morphogenesis, intestinal morphogenesis, neonatal development

INTRODUCTION

Increasingly, cell biologists are asking themselves questions not just about processes occurring within the bounds of the cellular membrane, but those that are happening outside of cells. We are learning more all the time about the complex ways in which the extracellular matrix (ECM) affects cell fate, behavior, and movement. The ECM is made up of proteins, glycoproteins, proteoglycans, and glycosaminoglycans which interact with each other and with receptors on the cell surface, (Lu, Weaver, and Werb 2012) and create a sieve-like network that affects the movement or sequestration of diffusible ligands through extracellular space (Rozario and DeSimone 2010; Bonnans, Chou, and Werb 2014; Dzamba and DeSimone 2018; Lieleg and Ribbeck 2011). The ECM is far from static. Cells sense their environment and respond to it, modulating the ECM by secreting structural, signaling, and enzymatic ECM components.

A developing embryo is incredibly dynamic, and thus the ECM of its tissues must also change to accommodate proper morphogenesis (Daley, Peters, and Larsen 2008). Rapid changes in embryo size, shape, and cellular distribution require a very soft ECM, which also helps to prevent stem cells from differentiating until they reach the appropriate location and timing to adopt their new fate (L. R. Smith, Cho, and Discher 2018). A key component of these soft ECM is hyaluronic acid (HA), which is essential for development from the fertilization of the egg until after birth. As the key functions of HA in fertility have been covered previously in this book, this chapter will focus on roles of HA and HA interactors during morphogenesis.

THE MANY FLAVORS OF HA

HA is a non-sulfated glycosaminoglycan found in diverse ECM, comprised of disaccharides of glucuronic acid and N-acetylglucosamine (Figure 1A) (Hascall and Esko 2017). HA can exist as very large molecules or very short oligos, which predictably has a drastic impact on its effects *in vivo*. Any HA molecule above 1000 kDa is considered to be high molecular weight HA (HMW-HA) and is described as protective to tissues: anti-proliferative, anti-inflammatory, and anti-cancer (Cowman et al. 2015; Slevin et al. 2007; Petrey and de la Motte 2014). HA within the range of 10 to 250 kDa is classified as low molecular weight (LMW) HA, which is pro-inflammatory and associated with some disease states including cancer and inflammation (Slevin et al. 2007; Petrey and de la Motte 2014). HA fragments smaller than 20 monosaccharides (oligo-HA/o-HA) are highly pro-angiogenic (Slevin et al. 2007) and can have a range of effects on inflammation depending on the context (Cyphert, Trempus, and Garantziotis 2015; Petrey and de la Motte 2014).

HA is synthesized at the cell membrane by hyaluronan synthase enzymes (HAS1, 2, and 3) (Figure 1B), which is one way in which the size of HA molecules is determined. HAS3 is known to synthesize shorter HA chains (100-1000 kDa) than HAS1 or HAS2 (Cyphert, Trempus, and Garantziotis 2015). HAS2, responsible for high molecular weight hyaluronan synthesis, is the predominant hyaluronan synthase in development; without it mouse embryos die by E10.5 from with serious defects including a lack of cardiac jelly that prevents proper heart morphogenesis (Camenisch et al. 2000). In contrast, Has1 (Kobayashi et al. 2010) and Has3 (Bai et al. 2005) knockout mice survive and appear phenotypically normal, as do Has1/3 double knockout mice (Mack et al. 2012).

HA size is also affected by hyaluronidases, which break HA molecules down (Figure 1B). The canonical hyaluronidase (Hyal) gene family contains four true hyaluronidases—Hyal1, Hyal2, and Hyal3 which catalyze HA in somatic tissues and Spam1 which has hyaluronidase activity limited to the testes—as well as Hyal4 (a chondroitinase, not a hyaluronidase) and Phyal1 (a pseudogene) (Hascall and Esko 2017). Different HYAL enzymes may break HA down to varying extents—for example, Hyal2 may break large HA molecules into smaller fragments, which are internalized and fully catabolized by HYAL1 in the lysosome (Figure 1B) (Hascall and Esko 2017). Although not a part of this canonical Hyal family, transmembrane protein 2 (Tmem2) and cell migration inducing hyaluronidase 1 (Cemip1/Kiaa1199) also play roles in HA depolymerization (Yamaguchi et al. 2019). We know that CEMIP1 is secreted and is important for HA degradation, but whether it has direct or indirect hyaluronidase activity remains unknown (Yoshida et al. 2013; Fink et al. 2015). TMEM2 is a membrane-bound hyaluronidase, existing as a single-pass transmembrane protein with its HA-cleaving domain outside the cell (Yamamoto et al. 2017). Tmem2 is of high interest in the field and its roles in development will be discussed later in this chapter.

HMW-HA can affect tissues by changing the mechanical properties of the tissue, where its hygroscopic nature causes tissue expansion. This is seen in a multitude of developmental contexts including the cumulus cell oocyte complex (COC) (Salustri et al. 1992) and cardiac cushions (Camenisch et al. 2000), and during the left-right looping morphogenesis of the intestine (Sivakumar et al. 2018) as will be described later. In some instances, this tissue expansion is further enhanced by the covalent addition of heavy chain proteins to the HA molecule by the enzyme TSG6, which stabilizes the HA in an expanded state (Figure 1B) (Day and Milner 2019). These heavy chain proteins come from the donor molecule inter-α-trypsin inhibitor (Iα) (Lauer et al. 2013). However, TSG6 may also condense an HA matrix by crosslinking the HA molecules (Day and Milner 2019). This means that TSG6 can fine-tune HA matrix expansion/condensation and thus tissue stiffness, consequently modulating mechanotransduction signaling pathways (Baranova et al. 2011; Humphrey, Dufresne, and Schwartz 2014). The relationship between TSG6 and HA is also important in contexts beyond the embryo including disease and therapeutics (reviewed by Day and Milner 2019).

There are several cell-surface receptors for HA that make it an effector in many genetic developmental programs (Figure 1B). Briefly, cluster of differentiation 44 (CD44) is a widely distributed cell-surface HA receptor which stimulates tyrosine kinases and Rho-like GTPase pathways (Turley, Noble, and Bourguignon 2002). Perhaps surprisingly, CD44-null embryos are viable and fertile without any severe phenotypes (Protin et al. 1999). Upon challenge, however, some roles of CD44 have been uncovered using CD44-null mice. In some cases, CD44 loss is a detriment: CD44-null mice over-respond to acute pulmonary inflammation from lipopolysaccharide (LPS) treatment (J. Liang et al. 2007) and have delayed wound healing and angiogenesis (Cao et al. 2006). In other cases, loss of CD44 offers a protective effect: for example, CD44-null mice are resistant to type 1 diabetes (Assayag-Asherie et al. 2015). The role of CD44 in cancer and inflammation is highly context-dependent, suggesting that other HA receptors and other ECM components that interact with CD44 (i.e., fibronectin, collagen, MMPs) may all contribute to these complex etiologies (Misra et al. 2015). Receptor for HA-mediated motility (RHAMM) has been described as localizing to the cell surface, nucleus, and cytoplasm of diverse cell types (Turley, Noble, and Bourguignon 2002), although its localization and function as an HA receptor are under debate (He et al. 2020). Specific to the lymphatic system, lymphatic vessel endothelial receptor-1 (LYVE1) is important for leukocyte docking to pull these immune cells into lymph vessels, as well as initiating signaling pathways (Jackson 2019). Interestingly, Lyve1-null mice are viable and do not appear to have any lymphatic defects (Gale et al. 2007). Other receptors include HA receptor for endocytosis (HARE) and layilin, which bind HA directly (Cyphert, Trempus, and Garantziotis 2015), and toll-like receptor 2 and 4, which bind HA indirectly (Ebid, Lichtnekert, and Anders 2014).

The complexity of HA interactions does not stop there, however. Hyaladherins are ECM components that bind HA, including the proteoglycans versican, aggrecan, brevican, and neurocan. These proteoglycans can be quite large, owing to a GAG binding domain and the attached GAGs chains which give these molecules a "bottle brush" appearance (Figure 1B) (Day and Prestwich 2002). Proteoglycans play important roles in many aspects of development, including that of the nervous system (reviewed by Howell and Gottschall 2012) with large versican isoforms and neurocan found at high levels in the late embryonic brain (Milev et al. 1998).

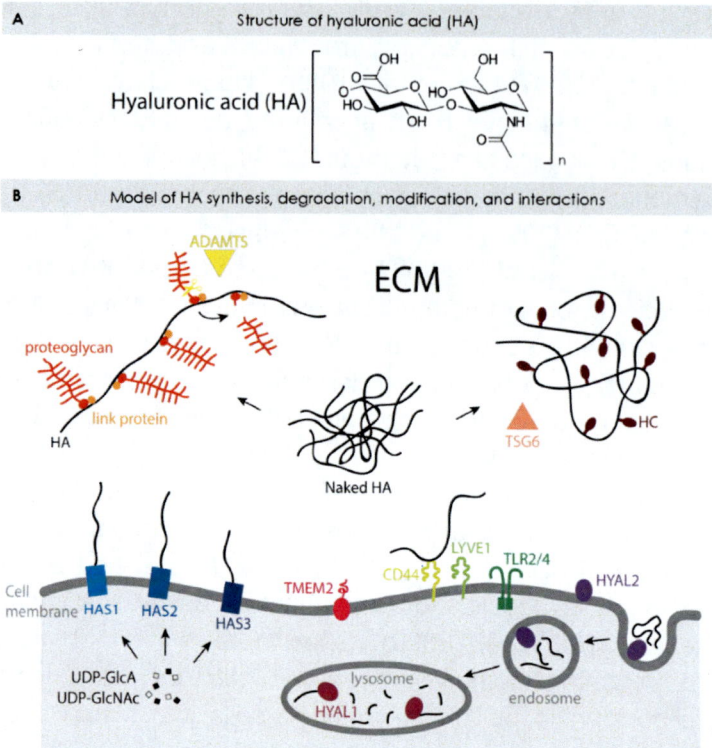

Figure 1. HA: a simple molecule with complex interactions. (A) Structure of Hyaluronic Acid (HA). (B) The synthesis, degradation, and interactions of HA. HA is synthesized into the extracellular space by Hyaluronan Synthase enzymes (HAS1, HAS2, and HAS3) from the subunits UDP-GlcA and UDP-GlcNAc. In the ECM, HA may be covalently modified by the enzyme TSG6, bound by proteoglycans (susceptible to cleavage by ECM enzymes including ADAMTSs and MMPs), and bound to cell-surface receptors (i.e., CD44, LYVE1, TLR2/4). HA can be broken down by hyaluronidases including HYAL2, TMEM2, and HYAL1.

These hyaladherins can be cleaved by some enzymes in the A Disintegrin and Metalloproteinase with Thrombospondin motifs (ADAMTS) or Matrix metalloproteinases (MMP) gene families (Apte 2013; Howell and Gottschall 2012). We may only understand the tip of the iceberg that is ADAMTS catalysis of proteoglycans, but their importance is already clear. Notably, ADAMTS9-null embryos die at the onset of gastrulation (Silver et al. 2008; Kern et al. 2010; Enomoto et al. 2010), and ADAMTS1 is required for full fertility (Russell, Brown, and Dunning 2015). Multiple developmental patterning processes depend on ADAMTS enzymes, including cardiac morphogenesis (Kern et al. 2006, 2007, 2010; Dupuis et al. 2011, 2013, 2019; Cooley et al. 2012; Stankunas et al. 2008), palatal fusion (Enomoto et al. 2010; Wolf et al. 2015), interdigital tissue death to prevent syndactyly (McCulloch et al. 2009), and melanocyte survival (Rao et al. 2003; Silver et al. 2008). Several of these examples will be expanded upon later.

We hope you can appreciate the incredible diversity of HA interactions with the ECM and cells that makes it a fascinating but challenging molecule to research. HA has been and continues to be studied extensively as a player in adult disease states, but studies on its influence in embryonic development have historically been limited to fertilization and heart morphogenesis. Here, we highlight some of the exciting work that has been done within the last five to ten years that brings to light the critical roles of HA in embryonic development.

HA IN CARDIAC MORPHOGENESIS

The study of HA in organogenesis began with the embryonic heart over two decades ago. Structures called cardiac cushions contain a soft, expanded ECM (termed "cardiac jelly"); these comprise most of the thickness of the heart wall at early stages (Barry 1948). Around E9.5, endocardial cells transition to mesenchymal cells and invade the cardiac jelly, beginning tissue remodeling that eventually molds the cardiac cushions into the valves and septa of the heart (recently reviewed by (Silva et al. 2021). During these critical early stages of heart development, versican and HA are co-expressed in the heart and are enriched in the cardiac jelly (Camenisch et al. 2000). Yamamura, Mjaatvedt, and colleagues demonstrated that versican is required for proper development of the mouse heart tube and the formation of the cardiac cushions (Yamamura et al. 1997; Mjaatvedt et al. 1998). Consequently, versican null mice (heart defect, or hdf mutant) die by E10.5 (Mjaatvedt et al. 1998). Has2 null mouse embryos, in which HMW-HA synthesis is abrogated, similarly lack cardiac jelly and have a condensed mesenchyme, so the cardiac cushions do not expand and the embryos die between E9.5 and E10.0 (Camenisch et al. 2000).

Importantly, this role of versican and HA is evolutionarily conserved far beyond the mammalian lineage. The Linear Heart Tube (lht) Medaka fish mutant has a mutation in the 3'UTR of the versican gene (effectively a versican knockdown) (Mittal et al. 2019). In these mutants, the heart does not loop to form two separate chambers, and cardiac cushions/jelly are absent (Mittal et al. 2019). In fish, cardiac jelly expands asymmetrically to properly form the septum and the inflow and outflow tracts (Derrick et al. 2021). Work in the zebrafish has shown that this asymmetry is not a consequence of uneven HA deposition. Instead, the asymmetric expansion is driven by the proteoglycan link protein 1a (HAPLN1a) (Derrick et al. 2021) which crosslinks proteoglycans and HA (Figure 1B). HAPLN1 is likely binding both versican and aggrecan, the latter of which has also been implicated in cardiac development for zebrafish, birds, and mammals (reviewed by Koch, Lee, and Apte 2020). Finally, the cleavage of versican and aggrecan by ADAMTS1, 5, and 9 has been shown to drive critical remodeling of the ECM in heart development (Kern et al. 2006, 2007, 2010; Dupuis et al. 2011, 2013, 2019; Cooley et al. 2012; Stankunas et al. 2008). This is exemplified by valve remodeling: while HA and versican cause expansion of the cardiac cushions, ADAMTS break down this ECM at later stages of development to condense the tissue into thin, tough valves that ensure efficient pumping of the heart (Kern et al. 2010, 2006; Dupuis et al. 2011, 2013, 2019).

During cardiac morphogenesis, the breakdown of HA molecules by hyaluronidases is as important as its synthesis, regulation, and binding interactions. Hyal2 null mice suffer a variety of maladies, including congenital heart defects. Hyal2 mutants have expanded heart valves and a disorganized ECM, and about half have atrial dilation (Chowdhury et al. 2013, 2017). Half of the embryos also have cor triatriatum (triatrial heart), a rare congenital defect where one atrium, often the left, is divided into two by a thin membrane (Chowdhury et al. 2016). Interestingly, human patients with Hyal2 mutations may also display cor triatriatum: a study of two human families with Hyal2 mutations was the first to provide a molecular explanation for this condition in humans (Muggenthaler et al. 2017). A more recently characterized hyaluronidase, TMEM2, has been determined to be an essential player in cardiac development as well. In 2011, two different groups published studies using independent zebrafish TMEM2 mutants picked from forward genetic screens (Totong et al. 2011; K. A. Smith et al. 2011). Both papers established that TMEM2 is important for proper morphogenesis of the atrioventricular canal (AVC) (Totong et al. 2011; K. A. Smith et al. 2011). The AVC separates the atrium and ventricle, and its proper formation is important for heart looping, septa formation, and valve formation to occur correctly (Peal, Lynch, and Milan 2011). Without TMEM2, markers of the AVC (such as Bmp4) are aberrantly expanded, suggesting that this hyaluronidase is necessary to restrict the AVC (Totong et al. 2011; K. A. Smith et al. 2011). As a consequence, the heart does not loop properly and embryos do not survive (Totong et al. 2011; K. A. Smith et al. 2011). When maternally-contributed TMEM2 protein is also removed, cardiac phenotypes became more severe (Totong et al. 2011). In particular, cardiac fusion—the meeting of cardiomyocytes and endocardial cells to form a closed heart tube—did not occur (Totong et al. 2011). A follow-up paper suggested that cardiac fusion failure may be a consequence of disorganized ECM in the TMEM2 mutants, preventing cell migration (Ryckebüsch et al. 2016). Alternatively, it is possible that loss of TMEM2 may negatively affect signaling via RHAMM or CD44, since these receptors have been shown to be important for myoblast migration and proliferation, respectively, in cultured embryonic mouse forelimbs (Leng et al. 2019). However, the true mechanism(s) of TMEM2 activity in embryonic development remains to be uncovered.

Cardiac morphogenesis is a fantastic example of the complex ways in which HA can affect organogenesis—at the level of synthesis, degradation, receptor binding, and proteoglycan binding—all of which must be carefully balanced. The carefully concerted expansion of the cardiac cushions and subsequent condensation to remodel into mature structures is arguably one of the most critical morphological processes. A failure at any level of HA regulation puts the organism at risk for embryonic lethality or congenital heart defects (seen in 0.4-2% of live births) (Miranović 2014). This beautiful complexity also applies to other developmental systems including the intestines, which will be discussed next.

HA IN INTESTINAL MORPHOGENESIS

Recently, a novel role for hyaluronan in chicken and mouse intestinal morphogenesis was identified by Sivakumar and colleagues (Sivakumar et al. 2018). Gut looping is a highly conserved process within a species—every individual should have the same looping patterns (Savin et al. 2011). Deviations from the prescribed looping pattern (intestinal malrotation) predisposes individuals for volvulus, a self-strangulation of the gut (Applegate 2009). The dorsal mesentery (DM) is a thin, left-right asymmetric organ that connects the gut tube to the rest of the body, and steers the gut to form the loops correctly (Figure 2A) (Kurpios et al. 2008; Davis et al. 2008). The gut begins as a straight tube aligned with the midline of the body, but asymmetric cellular and extracellular behavior of the DM drives the gut to tilt to the left. More specifically, the left mesenchyme of the DM condenses, while the right mesenchyme expands and the right epithelial cells transition from columnar to cuboidal morphology (Figure 2B) (Kurpios et al. 2008; Davis et al. 2008). This creates asymmetric forces that push the gut tube to the left, setting it up to form the first gut loop. At first, it was believed that the left side drove gut tilting because it is a Pitx2-positive compartment (Figure 2C) (Davis et al. 2008; Kurpios et al. 2008; Welsh et al. 2013), and this transcription factor is known to drive evolutionarily-conserved left-right organ asymmetry in other contexts (i.e., Ryan et al. 1998; Logan et al. 1998; Franco, Sedmera, and Lozano-Velasco 2017). However, careful time course experiments in the chick embryo by Sivakumar et al. showed that it is in fact the right side that initiates its expansion before the left side condenses (Figure 2B) (Sivakumar et al. 2018). This expansion is caused by an enrichment of HA on the right (Figure 2C) that occurs despite apparently symmetric expression of Has2 in the DM. This scenario is reminiscent of the zebrafish heart with asymmetric expansion driven by HAPLN1 (Derrick et al. 2021), except in the DM the enrichment is attributed to covalent modification of right-sided HA by TSG6 (Figure 2C, 3C) (Sivakumar et al. 2018). Without TSG6 or HA, the right side fails to expand and gut tilting does not occur (Figure 3A). The consequences of TSG6 loss was observed in mutant mice, as well, where gut looping patterns in E18.5 embryos were perturbed

(Sivakumar et al. 2018). This indicates that modification of HA by TSG6 is an evolutionarily-conserved mechanism critical for proper intestinal morphogenesis. Importantly, to date this is the only known function of TSG6 in embryonic development besides COC expansion.

The role of HA in the DM extends beyond gut tilting, however. As noted, the left side also synthesizes HA, but it is not modified by TSG6 and is not sufficient for expansion (Sivakumar et al. 2018). Another key difference between the left and right HA is its effect on vasculature. Prior to right-sided expansion and gut tilting, there are two lines of endothelial cells in the DM, one on each side, with the capacity for differentiating into blood vessels. However, as the right side expands, the right-sided endothelial cells disperse, leaving only the left-sided endothelial cells to form a key gut artery (Figure 2B) (Sivakumar et al. 2018). This artery later branches and goes on to form the ileocolic and middle colic arteries which supply a large portion of the adult intestine with blood. Interestingly, Sivakumar et al. showed that both the exclusion of endothelial cells from the right and the formation of the artery on the left depend on the respective "flavors" of HA on each side. That is, TSG6-modified HA on the right is necessarily anti-angiogenic (Figure 3B), and the unmodified HA on the left is necessarily pro-angiogenic (Figure 3C) (Sivakumar et al. 2018). Whether HA size or the involvement of other HA interactors contributes to these opposing characteristics is currently being studied.

HA IN ANGIOGENESIS

Vascular development is complex and may be driven by a variety of pathways including vascular endothelial growth factor (VEGF), angiopoietin, platelet-derived growth factor, transforming growth factor β, bone morphogenetic protein, Notch, and other signaling pathways (Grant and Coultas 2019). HA can affect angiogenesis by initiating signaling cascades through its receptors CD44 and RHAMM (Pardue, Ibrahim, and Ramamurthi 2008). As noted previously, the size of HA molecules present in a tissue has significant influence on its effects on angiogenesis. HA was first implicated in vasculogenesis by Richard Feinberg and David Beebe in 1983, who made the observation that HA isolated from human umbilical cords is anti-angiogenic when introduced to chick limb buds (Feinberg and Beebe 1983). This was later determined to be HMW-HA (Kanayama, Goto, and Terao 1999). Conversely, the highly active area of study of HA in cancer biology has identified oligo-HA as a positive regulator of tumor vascularization (Pardue, Ibrahim, and Ramamurthi 2008).

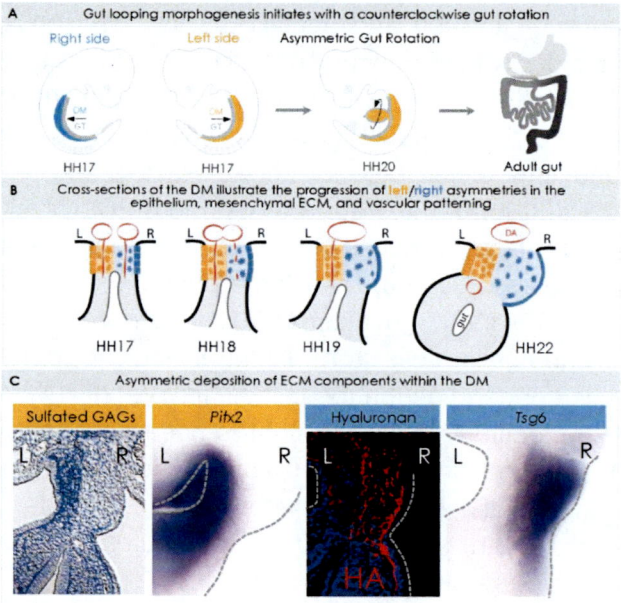

Figure 2. The asymmetric dorsal mesentery (DM) drives asymmetric gut rotation and vascular patterning. (A) The proper establishment of the first gut loop around chicken development stage 20 (Hamburger-Hamilton stage 20, HH20) is critical for setting up subsequent looping patterns to make the highly stereotypical adult intestinal looping pattern. (B) The DM is cellularly symmetrical at early stages (HH17). The right side begins its expansion (HH18) before the left side condenses (HH19). Vascular precursor cells (red) are also symmetrical at HH17, with a line of cells on both the right and left sides. As the right side initiates its expansion, driving the gut tube leftward, the right side also excludes its endothelial cells, resulting in a left-sided gut artery (HH22). (C) Molecular asymmetries in the DM are observed with transcription factors (Pitx2) and also with ECM components including glycosaminoglycans (GAGs), HA, and TSG6 (data from chicken embryos). Figure modified from Sivakumar et al., 2018.

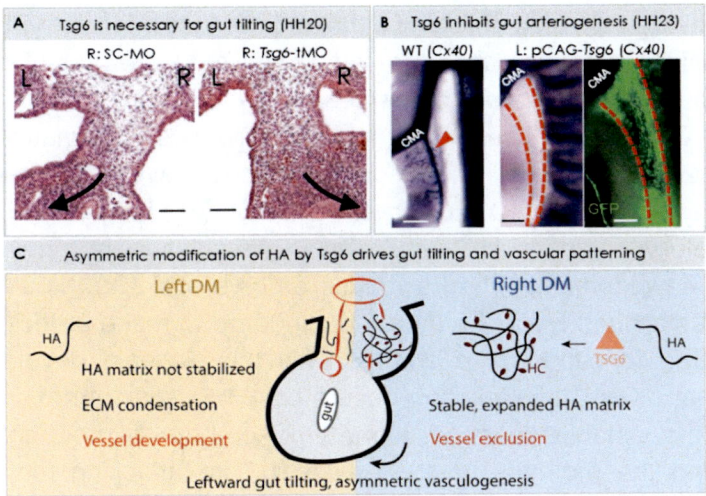

Figure 3. Modification of HA by TSG6 is necessary for gut tilting and proper vascular patterning. (A) When Tsg6 is knocked down in the chick using a translation-blocking morpholino (tMO), the leftward gut tilting seen in scrambled morpholino (SC-MO) controls is lost. (B) RNA *in situ* hybridization for the vascular marker Cx40. Ectopic expression of Tsg6 on the left side of the DM results in a loss of a key gut artery (red arrowhead) that normally branches off of the cranial mesenteric artery (CMA). GFP is co-electroporated with Tsg6 to visualize the tissue targeted (DM, red dashed lines). (C) Model of HA and its effects in the DM. TSG6-modified HA on the right forms a stable matrix with anti-vascular properties, while unmodified HA on the left is necessary for vascular development. Figure modified from Sivakumar et al., 2018.

In the earlier example of arterial patterning in the intestine, the Cxcl12/Cxcr4 pathway is critical (Sivakumar et al. 2018). When the DM is symmetrical, the mRNA expression of the Cxcl12 chemokine is bilateral, but it becomes exclusively left-sided as asymmetries are established (Sivakumar et al. 2018). HA is a negative transcriptional regulator of Cxcl12, the ligand for the G protein-coupled chemokine receptor Cxcr4 (Mueller, Yoon, and Sadiq 2014; Sivakumar et al. 2018). It follows, then, that a loss of HA from the right chick DM results in the persistence of bilateral Cxcl12 expression (Sivakumar et al. 2018). VEGF signaling may also be involved, as unpublished data from the Kurpios group shows that VEGFA is expressed in the left DM mesenchyme and VEGFR2 is expressed by the left-sided endothelial cells. Although the size of HA on either side of the DM is not yet known, the possibility remains that the left could have LMW HA or o-HA affording it pro-angiogenic properties, while the right side may have HMW-HA that could sequester signaling ligands. Alternatively, this regulation could occur at the transcriptional level (HA negatively regulates Cxcl12, Cxcl12 is upstream of VEGF (Z. Liang et al. 2007).

Carefully balanced HA regulation in vasculature is required very early in development. When a mouse embryo sticks to the surface of the uterus around E4.0, the uterus is triggered to make a decidua with vasculature to support the embryo until the placenta develops fully (Dey et al. 2004). Implantation occurs by E4.5, and with this process comes the establishment of a maternal-embryo barrier, critical for preventing the mother's immune system from attacking the embryo (Mori et al. 2016). The thickness of this maternal-embryo barrier is important: if it is too thick, diffusion of gases, nutrients, and wastes will be limited, if it is too thin, there is a risk that maternal and fetal blood could mix and cause complications in pregnancy (Hadas et al. 2020). New work from Michal Neeman's group describes the necessity of HA for vascular remodeling upon implantation into the mouse uterus and controlling the thickness of the maternal-embryo barrier (Hadas et al. 2020). High molecular weight hyaluronan acts as a negative angiogenic morphogen during early pregnancy. In contrast, hyaluronan breakdown products, generated upon degradation by hyaluronidases, promote vascular permeability via the VEGF-VEGFR2 signaling pathway to support perfusion to the developing embryo. Genetic manipulations evidenced that overexpressing Hyal2 to increase HA cleavage in the trophoblast impairs implantation by inducing decidual permeability, causing defects in the vascular sinuous folds (VSFs), and breaching the maternal-embryo barrier. In contrast, enhanced deposition of hyaluronan in the post-implantation period (through Has2 overexpression—normally Has2 expression decreases during this period) resulted in the thickening of the maternal-embryo barrier and increased diffusion distance (Hadas et al. 2020).

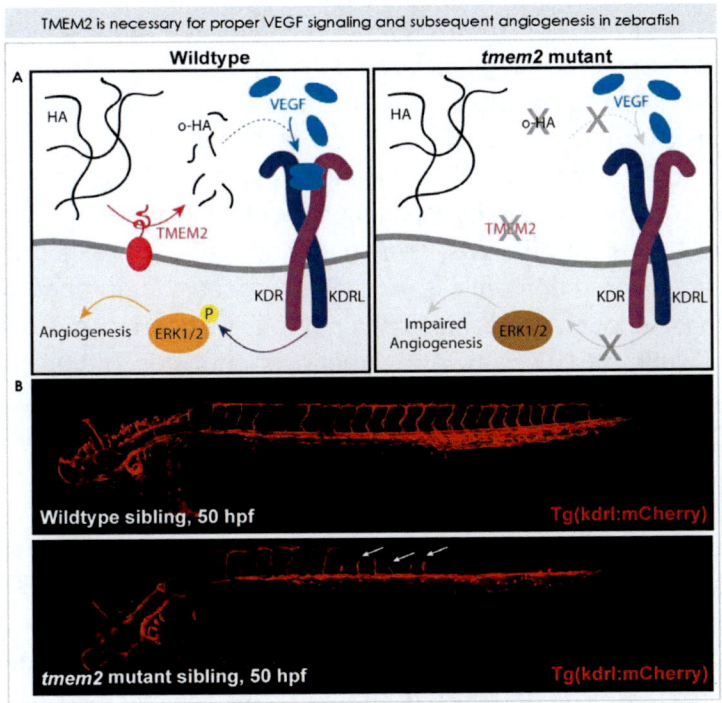

Figure 4. Catalysis of HA by TMEM2 is necessary for healthy vascular development in the zebrafish. (A) Model of HA processing by TMEM2 and subsequent signaling to stimulate angiogenesis in the zebrafish embryo, comparing pathways in the wildtype and tmem2 mutant. Although the mechanism is not clear, o-HA facilitates VEGF signaling through the Kdr/Kdrl receptor. (B) Sibling zebrafish embryos at 50 hours post-fertilization (hpf), with the receptor Kdrl labeled with mCherry to visualize blood vasculature. TMEM2 mutants (bottom) display incomplete or absent intersegmental vessels (ISVs, white arrows), contrasting a wildtype sibling control (top). Figure modified from De Angelis et al., *Developmental Cell* 2017.

De Angelis et al. (2017) highlighted the effect of TMEM2 on VEGF signaling and its necessity for healthy vascular development in the zebrafish. In the zebrafish, VEGF ligands can activate multiple downstream pathways through the VEGFR2/KDR receptor, most notably pERK1/2 signaling for angiogenesis (Figure 4A) (Shin, Male, et al. 2016; Shin, Beane, et al. 2016). Normally, HA localizes neatly to the large blood vessels (i.e., the dorsal aorta) in the zebrafish. In TMEM2 mutants however, HA aberrantly collected in surrounding tissues as well. These mutants also exhibited a striking decrease in the number of intersegmental vessels (ISVs) that grow in the embryo's truck (Figure 4B). The ISVs were able to sprout off of the existing vessel, but not continue their outgrowth. The over-accumulation of HA and vascular defects could be rescued by injecting either hyaluronidase or o-HA into the TMEM2 mutant embryos, demonstrating that the breakdown of HA into these small fragments is a critical process for proper vascular patterning. Although a mechanism is not totally understood, the authors implicated the VEGF pathway in these phenotypes (Figure 4A) (De Angelis et al. 2017). It remains under debate whether TMEM2 affects the VEGF pathway transcriptionally or at the level of ligand/receptor binding (De Angelis et al. 2017; Rodgers et al. 2006).

As with other morphogenetic processes, "friends of HA" (binding partners and modifiers) play important roles in angiogenesis. The same study showing the importance of versican in medaka fish heart development found that versican is also necessary for functional blood vessel formation (Mittal et al. 2019). Regardless of whether the vessel arose from vasculogenesis (formation of new vessel from cord hollowing) or angiogenesis (branching from existing vessel), vessels in the lht mutant (versican knockdown) embryos lacked a lumen and thus these embryos had no blood circulation (Mittal et al. 2019). Versican and HA have also been implicated in the proper development of yolk sac vasculature by Suneel Apte's group. The *de novo* development of blood vessels in the yolk sac did not occur in versican-null hdf mice, and HA was absent from the yolk sac mesoderm (Nandadasa et al. 2021). Has1/2/3 triple knockout mice also have avascular yolk sacs and conversely lose versican in the yolk sac and embryo (Nandadasa et al. 2021). However, cleaved versican is still present in the embryo, suggesting that HA might shield versican from cleavage by ADAMTS enzymes (Nandadasa et al. 2021). The authors suggest that these phenotypes may arise from migration defects in Flk1+ hematoendotheial progenitor cells, which migrate from the primitive streak and form blood islands in the yolk sac around E7. Alternatively, versican may sequester pro-angiogenic factors like VEGF-A to support extra-embryonic vascular development (Nandadasa et al. 2021). As a final example of modified HA playing a role in angiogenesis, Sivakumar et al. showed that covalent modification of HA by TSG6 is required for vascular exclusion in the right chick DM. When TSG6 is ectopically expressed on the left, the left-sided artery does not form (Figure 3B) (Sivakumar et al. 2018).

The body of research on the role of HA in angiogenesis spans animal classes—fish, birds, and mammals—and cooperatively demonstrates the critical, conserved need for proper HA in blood vasculature development. To delve deeper into this topic, refer to a review by Pardue, Ibrahim, and Ramamurthi (2008).

HA IN CRANIOFACIAL AND SKELETAL DEVELOPMENT AND ODONTOGENESIS

Craniofacial development heavily depends on correct migration of cranial neural crest cells (NCCs). In 1975, biochemical analyses of the cell-free spaces through which chick neural crest cells migrate showed that the ECM is highly enriched for HA (Pratt, Larsen, and Johnston 1975). Cranial NCCs in *Xenopus laevis* embryos express Has1 and CD44, while the tissue through which they migrate (the pharyngeal endoderm) expresses Has2 (Casini, Nardi, and Ori 2012). The functional activities of these genes were studied using antisense morpholinos (MO) to knock down Has1, Has2, and CD44 in one blastomere at the two- or four-cell stage. Has1-MO and Has2-MO both caused delays in NCC migration and increased apoptosis in post-migratory cells, resulting in craniofacial abnormalities ranging from mild (all facial structures present, but diminished in size) to severe (some or all structures absent or much smaller than the control side of the embryo) (Casini, Nardi, and Ori 2012). CD44-MO injected embryos also had delays in NCC migration, although fewer had craniofacial defects (Casini, Nardi, and Ori 2012). Collectively, these data suggest that HA is a pro-migratory ECM component for NCC migration. Versican, on the other hand, has been identified as a barrier against NCC migration (Landolt et al. 1995). This proteoglycan has been shown *in vitro* to act as a guide for NCCs, as a fence guides a flock of moving sheep (Dutt et al. 2006).

Cleft lip with or without cleft palate is one of the most common human congenital birth defects. Closure of the secondary palate is highly dependent on a dynamic ECM (reviewed in Wang et al. 2020), including HA and versican. When Has2 is conditionally knocked out in cranial NCCs, mutant mice have cleft palate due to diminished palatal shelves that do not move and fuse properly (Yonemitsu, Lin, and Yu 2020). Mutant neonates display micrognathia (underdeveloped lower jaw), tongue protrusion, and complete cleft palate, resulting in survival times of only a few hours because they struggle to breathe and nurse (Yonemitsu, Lin, and Yu 2020). Interestingly, one third of Hyal2-null mice also develop cleft lip/palate that prevents feeding, leading to perinatal death (Muggenthaler et al. 2017). This phenotype is reflected in humans that carry Hyal2 mutations (Muggenthaler et al. 2017). Versican and its cleavage by ADAMTS9 and ADAMTS20 has also been shown to be essential in palate closure (Enomoto et al. 2010). ADAMTS9 +/-, bt/bt ("belted," ADAMTS20 -/-) mice have fully penetrant cleft palate, and 65% of bt/bt, hdf/+ (ADAMTS20 null, versican het) mice have cleft palate as well (Enomoto et al. 2010). The authors attribute this phenotype to a decrease in mesenchymal cell proliferation (Enomoto et al. 2010). Genome-wide association studies (GWAS) connected mutations in ADAMTS20 with cleft lip/palate in humans and dogs (Wolf et al. 2015), further supporting the important role of ECM remodeling in craniofacial development.

The hyaladherin aggrecan is well-established as a critical player in skeletal development throughout the body. In 1965, Walter Landauer published an account of the recessive "nanomelia" mutation in chick embryos, describing severe spinal defects, shortened limbs, splayed legs, and a misshapen beak (Landauer 1965). This was later determined to be a mutation in the aggrecan gene (Li, Schwartz, and Vertel 1993; Primorac et al. 1994; Vertel et al. 1994). Aggrecan-null mice (cartilage matrix deficiency model, or cmd) are much smaller than heterozygous littermates, and have a severely underdeveloped skeletal system due to a lack of cartilage matrix. Craniofacial defects in mutant embryos include a short snout, cleft palate, and a protruding tongue, which likely contribute to the perinatal lethality from respiratory failure (Watanabe et al. 1994). While cmd heterozygotes appear normal at birth, they develop dwarfism and spinal misalignments leading to degeneration within the first year of life. This leads to a loss of mobility that prevents feeding; consequently, these animals die from starvation when they are only 12-15 months old, with no heterozygotes living longer than 19 months (Watanabe et al. 1997; Watanabe and Yamada 2002). The essential cooperation of HA and aggrecan in skeletal development is supported by two pieces of evidence. First, a limb mesenchyme-specific Has2 knockout mouse model (*Prx-1 Cre, Has2 fl/fl*) produced mutants with shortened limb bones, defective joints, and mispatterned phalanges (Matsumoto et al. 2009). Second, link protein knockout mouse mutants show moderate dwarfism, shortened long bones, and mild craniofacial defects, with most homozygotes suffering perinatal lethality from respiratory failure (Watanabe and Yamada 2002). Collectively, these phenotypes highlight the important partnership between HA and aggrecan in skeletal development.

Odontogenesis (tooth development) also uses HA to regulate proliferation. Molar development happens similarly in mice and humans, with three molars budding off of one molar dental placode, a thickening of the oral epithelium (Lumsden 1979). When E14.5 mouse dental placodes are cultured and exposed to 4-methylumbelliferone (4-MU, an inhibitor of HA synthesis), more proliferation occurs and the first molar (M1) is larger than in controls (Sánchez et al. 2020). However, subsequent molars do not bud off of M1 during the limited time of culturing, while control placodes produce two molars (M1 and M2) and have a tail where M3 would form. Thus, HA has an influence on both tooth size and tooth number (Sánchez et al. 2020).

HA IN NEONATAL DEVELOPMENT

Given the critical roles for HA in embryonic development and in adult tissues, it should come as no surprise that HA is an essential player in neonatal development as well. In this delicate period of life, neonates must adjust to breathing and eating for themselves. While post-natal development of the digestive system will be discussed in the next chapter, here we will briefly outline what is known about the role of HA in the post-natal respiratory system.

In humans (Johnsson et al. 2003) and other mammals (Allen et al. 1991), the amount of HA in the lung decreases as the fetus approaches birth. HA helps to hold water in the lung tissue. This could partially explain why prematurely-born babies often have respiratory syndromes (Fraser, Walls, and McGuire 2004)—aberrantly high HA at birth means greater hydration of lung tissues (Johnsson et al. 1998). Recently, a negative correlation was observed between RHAMM protein levels and air space in the lung of deceased newborn patients diagnosed with respiratory conditions (Markasz et al. 2018). In a healthy newborn or adult lung, RHAMM levels are low, but the researchers here suggest that in neonatal respiratory pathologies, RHAMM levels may be elevated. In contrast, alveolar macrophages in the fetal lung are enriched for CD44, and are important for reducing HA levels (and thus water content) in the lung tissue at birth (Martine Culty, Nguyen, and Underhill 1992; Underhill et al. 1993; M. Culty et al. 1994; Johnson et al. 2018). The body of research on HA in the developing and neonatal lung has a lot of potential for expansion in the coming years.

Conclusion

The dynamism of the developing embryo depends on a dynamic ECM. The works summarized here illustrate the critical role that HA synthases, hyaluronidases, receptors, hyaladherins, and modifiers have on a variety of developmental processes, from early events like embryo implantation to post-natal acclimation to breathing and feeding (Figure 5). HA may appear to be a simple sugar, but its effects on vertebrate life are fascinatingly complex. There is no doubt that the coming decades will be filled with further discovery in this field.

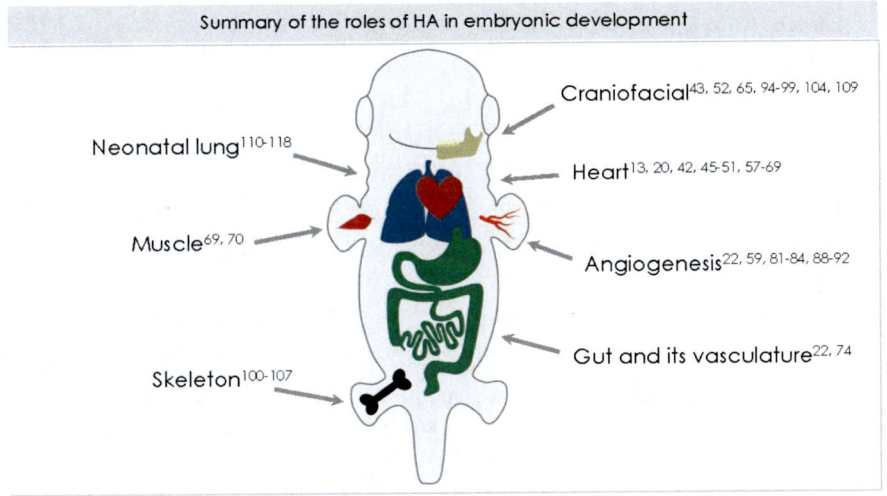

Figure 5. Summary of the roles of HA discussed in this chapter, with key references.

ACKNOWLEDGMENTS

We thank Drs. Kelly Smith and Anne Lagendijk for generously helping with the creation of Figure 4 in this chapter. Our appreciation goes to the HA field for invigorating discussions about the exciting HA research being done around the world, and we apologize to anyone whose work was not highlighted in this chapter. This work was supported by the National Science Foundation (NSF GRFP 2018257595) (C.M.D.); March of Dimes 1-FY11-520 (N.A.K.), and NIDDK R01 DK092776 and DK107634 (N.A.K.).

REFERENCES

Allen, S. J., Sedin, E. G., Jonzon, A., Wells, A. F. & Laurent, T. C. (1991). "Lung Hyaluronan during Development: A Quantitative and Morphological Study." *American Journal of Physiology - Heart and*

Circulatory Physiology, 260, H1449–54. https://doi.org/10.1152/ajpheart.1991.260.5.h1449.

Applegate, Kimberly E. (2009). "Evidence-Based Diagnosis of Malrotation and Volvulus." Pediatric Radiology, 39, 161–63. https://doi.org/10.1007/s00247-009-1177-x.

Apte, Suneel. (2013). Chapter 259 The ADAMTS Endopeptidases. Handbook of Proteolytic Enzymes. Vol. 1. https://doi.org/10.1016/B978-0-12-382219-2.00258-1.

Assayag-Asherie, Nathalie, Dror Sever, Marika Bogdani, Pamela Johnson, Talya Weiss, Ariel Ginzberg, Sharon Perles., et al. (2015). "Can CD44 Be a Mediator of Cell Destruction? The Challenge of Type 1 Diabetes." PLoS ONE, 10 (12). https://doi.org/10.1371/journal.pone.0143589.

Bai, Kuan Jen, Andrew P. Spicer, Marcella M. Mascarenhas, Lunyin Yu, Cristhiaan D. Ochoa, Hari G. Garg. & Deborah A. Quinn. (2005). "The Role of Hyaluronan Synthase 3 in Ventilator-Induced Lung Injury." American Journal of Respiratory and Critical Care Medicine, 172 (1). https://doi.org/10.1164/rccm.200405-652OC.

Baranova, Natalia S., Erik Nilebäck, F. Michael Haller, David C. Briggs, Sofia Svedhem, Anthony J. Day. & Ralf P. Richter. (2011). "The Inflammation-Associated Protein TSG-6 Cross-Links Hyaluronan via Hyaluronan-Induced TSG-6 Oligomers." Journal of Biological Chemistry, 286 (29), 25675–86. https://doi.org/10.1074/jbc.M111.247395.

Barry, Alexander. (1948). "The Functional Significance of the Cardiac Jelly in the Tubular Heart of the Chick Embryo." The Anatomical Record, 102 (3). https://doi.org/10.1002/ar.1091020304.

Bonnans, Caroline, Jonathan Chou. & Zena Werb. (2014). "Remodelling the Extracellular Matrix in Development and Disease." Nature Reviews Molecular Cell Biology, 15 (12), 786–801. https://doi.org/10.1038/nrm3904.

Camenisch, Todd D., Andrew P. Spicer, Tammy Brehm-Gibson, Jennifer Biesterfeldt, Mary Lou Augustine, Anthony Calabro, Steven Kubalak, Scott E. Klewer. & John A. McDonald. (2000). "Disruption of Hyaluronan Synthase-2 Abrogates Normal Cardiac Morphogenesis and Hyaluronan-Mediated Transformation of

Epithelium to Mesenchyme." *Journal of Clinical Investigation*, 106 (3). 349–60. https://doi.org/10.1172/JCI10272.

Cao, Gaoyuan, Rashmin C. Savani, Melane Fehrenbach, Chris Lyons, Lin Zhang, George Coukos. & Horace M. DeLisser. (2006). "Involvement of Endothelial CD44 during *in Vivo* Angiogenesis." *American Journal of Pathology*, 169 (1). https://doi.org/10.2353/ajpath.2006.060206.

Casini, Paola, Irma Nardi. & Michela Ori. (2012). "Hyaluronan Is Required for Cranial Neural Crest Cells Migration and Craniofacial Development." *Developmental Dynamics*. https://doi.org/10.1002/dvdy.23715.

Chowdhury, Biswajit, Richard Hemming, Sabine Hombach-Klonisch, Bruno Flamion. & Barbara Triggs-Raine. (2013). "Murine Hyaluronidase 2 Deficiency Results in Extracellular Hyaluronan Accumulation and Severe Cardiopulmonary Dysfunction." *Journal of Biological Chemistry*, 288 (1), 520–528. https://doi.org/10.1074/jbc.M112.393629.

Chowdhury, Biswajit, Bo Xiang, Michelle Liu, Richard Hemming, Vernon W. Dolinsky. & Barbara Triggs-Raine. (2017). "Hyaluronidase 2 Deficiency Causes Increased Mesenchymal Cells, Congenital Heart Defects, and Heart Failure." *Circulation: Cardiovascular Genetics*, 10 (1). https://doi.org/10.1161/CIRCGENETICS.116.001598.

Chowdhury, Biswajit, Bo Xiang, Martina Muggenthaler, Vernon W. Dolinsky. & Barbara Triggs-Raine. (2016). "Hyaluronidase 2 Deficiency Is a Molecular Cause of Cor Triatriatum Sinister in Mice." *International Journal of Cardiology*, 209, 281–83. https://doi.org/10.1016/j.ijcard.2016.02.072.

Cooley, Marion A., Victor M. Fresco, Margaret E. Dorlon, Waleed O. Twal, Nathan V. Lee, Jeremy L. Barth, Christine B. Kern, Luisa Iruela-Arispe, M. & Scott Argraves, W. (2012). "Fibulin-1 Is Required during Cardiac Ventricular Morphogenesis for Versican Cleavage, Suppression of ErbB2 and Erk1/2 Activation, and to Attenuate Trabecular Cardiomyocyte Proliferation." *Developmental Dynamics*, 241, 303–14. https://doi.org/10.1002/dvdy.23716.

Cowman, Mary K., Hong Gee Lee, Kathryn L. Schwertfeger, James B. McCarthy. & Eva A. Turley. (2015). "The Content and Size of

Hyaluronan in Biological Fluids and Tissues." *Frontiers in Immunology*, 6. https://doi.org/10.3389/fimmu.2015.00261.

Culty, M., O'Mara, T. E., Underhill, C. B., Yeager, H. & Swartz, R. P. (1994). "Hyaluronan Receptor (CD44) Expression and Function in Human Peripheral Blood Monocytes and Alveolar Macrophages." *Journal of Leukocyte Biology*, 56 (5). https://doi.org/10.1002/jlb.56.5.605.

Culty, Martine, Huong A. Nguyen. & Charles B. Underhill. (1992). "The Hyaluronan Receptor (CD44) Participates in the Uptake and Degradation of Hyaluronan." *Journal of Cell Biology*, 116 (4). https://doi.org/10.1083/jcb.116.4.1055.

Cyphert, Jaime M., Carol S. Trempus. & Stavros Garantziotis. (2015). "Size Matters: Molecular Weight Specificity of Hyaluronan Effects in Cell Biology." *International Journal of Cell Biology*. https://doi.org/10.1155/2015/563818.

Daley, William P., Sarah B. Peters. & Melinda Larsen. (2008). "Extracellular Matrix Dynamics in Development and Regenerative Medicine." *Journal of Cell Science*, 121, 255–64. https://doi.org/10.1242/jcs.006064.

Davis, Nicole M., Natasza A Kurpios, Xiaoxia Sun, Jerome Gros, James F Martin. & Clifford J Tabin. (2008). "The Chirality of Gut Rotation Derives from Left-Right Asymmetric Changes in the Architecture of the Dorsal Mesentery." *Developmental Cell*, 15 (1), 134–45. https://doi.org/10.1016/j.devcel.2008.05.001.

Day, Anthony J. & Caroline M. Milner. (2019). "TSG-6: A Multifunctional Protein with Anti-Inflammatory and Tissue-Protective Properties." *Matrix Biology*, 78–79, (January), 60–83. https://doi.org/10.1016/j.matbio.2018.01.011.

Day, Anthony J. & Glenn D. Prestwich. (2002). "Hyaluronan-Binding Proteins: Tying up the Giant." *Journal of Biological Chemistry*, 277 (7), 4585–88. https://doi.org/10.1074/jbc.R100036200.

De Angelis, Jessica E., Anne K. Lagendijk, Huijun Chen, Alisha Tromp, Neil I. Bower, Kathryn A. Tunny, Andrew J. Brooks, et al. (2017). "Tmem2 Regulates Embryonic Vegf Signaling by Controlling Hyaluronic Acid Turnover." *Developmental Cell*, 40, 123–36. https://doi.org/10.1016/j.devcel.2016.12.017.

Derrick, Christopher J., Juliana Sánchez-Posada, Farah Hussein, Federico Tessadori, Eric JG Pollitt, Aaron M Savage, Robert N Wilkinson, et al. (2021). "Asymmetric Hapln1a Drives Regionalised Cardiac ECM Expansion and Promotes Heart Morphogenesis in Zebrafish Development." *Cardiovascular Research*. https://doi.org/ https://doi.org/10.1093/cvr/cvab004.

Dey, S. K., Lim, H., Sanjoy K. Das, Jeff Reese, Paria, B. C., Takiko Daikoku. & Haibin Wang. (2004). "Molecular Cues to Implantation." *Endocrine Reviews*. https://doi.org/10.1210/er.2003-0020.

Dupuis, Loren E., Daniel R. McCulloch, Jessica D. McGarity, Alexandria Bahan, Andy Wessels, Deidra Weber, A. Megan Diminich, Courtney M. Nelson, Suneel S. Apte. & Christine B. Kern. (2011). "Altered Versican Cleavage in ADAMTS5 Deficient Mice; A Novel Etiology of Myxomatous Valve Disease." *Developmental Biology*, 357 (1), 152–64. https://doi.org/10.1016/j.ydbio.2011.06.041.

Dupuis, Loren E., Lockett Nelson, E., Brittany Hozik, Sarah C. Porto, Alexandra Rogers-DeCotes, Amanda Fosang. & Christine B. Kern. (2019). "Adamts5-/- Mice Exhibit Altered Aggrecan Proteolytic Profiles That Correlate With Ascending Aortic Anomalies." *Arteriosclerosis, Thrombosis, and Vascular Biology*, 39 (10), 2067–81. https://doi.org/10.1161/ATVBAHA.119.313077.

Dupuis, Loren E., Hanna Osinska, Michael B. Weinstein, Robert B. Hinton. & Christine B. Kern. (2013). "Insufficient Versican Cleavage and Smad2 Phosphorylation Results in Bicuspid Aortic and Pulmonary Valves." *Journal of Molecular and Cellular Cardiology*, 60, 50–59. https://doi.org/10.1016/j.yjmcc.2013.03.010.

Dutt, Shilpee, Maurice Kléber, Mattia Matasci, Lukas Sommer. & Dieter R. Zimmermann. (2006). "Versican V0 and V1 Guide Migratory Neural Crest Cells." *Journal of Biological Chemistry*, 281 (17). https://doi.org/10.1074/jbc.M510834200.

Dzamba, Bette J. & Douglas W. DeSimone. (2018). "Extracellular Matrix (ECM) and the Sculpting of Embryonic Tissues." In *Current Topics in Developmental Biology*, 245–74. https://doi.org/10.1016/bs.ctdb.2018.03.006.

Ebid, Rainer, Julia Lichtnekert. & Hans-Joachim Anders. (2014). "Hyaluronan Is Not a Ligand but a Regulator of Toll-Like Receptor

Signaling in Mesangial Cells: Role of Extracellular Matrix in Innate Immunity." *ISRN Nephrology.* https://doi.org/10.1155/2014/714081.

Enomoto, H., Nelson, C. M., Somerville, R. P. T., Mielke, K., Dixon, L. J., Powell, K. & Apte, S. S. (2010). "Cooperation of Two ADAMTS Metalloproteases in Closure of the Mouse Palate Identifies a Requirement for Versican Proteolysis in Regulating Palatal Mesenchyme Proliferation." *Development,* 137, 4029–38. https://doi.org/10.1242/dev.050591.

Feinberg, Richard N. & David C. Beebe. (1983). "Hyaluronate in Vasculogenesis." *Science,* 220 (4602), 1177–79. https://doi.org/10.1126/science.6857242.

Fink, Stephen P, Lois L Myeroff, Revital Kariv, Petra Platzer, Baozhong Xin, Debra Mikkola, Earl Lawrence, et al. (2015). "Induction of KIAA1199/CEMIP Is Associated with Colon Cancer Phenotype and Poor Patient Survival." *Oncotarget,* 6 (31), 30500–515. https://doi.org/10.18632/oncotarget.5921.

Franco, Diego, David Sedmera. & Estefanía Lozano-Velasco. (2017). "Multiple Roles of Pitx2 in Cardiac Development and Disease." *Journal of Cardiovascular Development and Disease,* 4 (4). https://doi.org/10.3390/jcdd4040016.

Fraser, Jenny, Moira Walls. & William McGuire. (2004). "ABC of Preterm Birth: Respiratory Complications of Preterm Birth." *British Medical Journal,* 329 (7472), 962–65. https://doi.org/10.1136/sbmj.0507278.

Gale, Nicholas W., Remko Prevo, Jorge Espinosa, David J. Ferguson, Melissa G. Dominguez, George D. Yancopoulos, Gavin Thurston. & David G. Jackson. (2007). "Normal Lymphatic Development and Function in Mice Deficient for the Lymphatic Hyaluronan Receptor LYVE-1." *Molecular and Cellular Biology,* 27 (2). https://doi.org/10.1128/mcb.01503-06.

Grant, Zoe L. & Leigh Coultas. (2019). "Growth Factor Signaling Pathways in Vascular Development and Disease." *Growth Factors,* 37, 53–67. https://doi.org/10.1080/08977194.2019.1635591.

Hadas, Ron, Eran Gershon, Aviad Cohen, Ofir Atrakchi, Shlomi Lazar, Ofra Golani, Bareket Dassa, et al. (2020). "Hyaluronan Control of the Primary Vascular Barrier during Early Mouse Pregnancy Is

Mediated by Uterine NK Cells." *JCI Insight*, 5 (22). https://doi.org/10.1172/jci.insight.135775.

Hascall, Vincent. & Jeffrey D. Esko. (2017). "Hyaluronan." In *Essentials of Glycobiology*, edited by A Varki, RD Cummings, and JD Esko, 3rd ed. Cold Spring Harbor (NY): Cold Spring Harbor Laboratory Press.

He, Zhengcheng, Lin Mei, Marisa Connell. & Christopher A. Maxwell. (2020). "Hyaluronan Mediated Motility Receptor (HMMR) Encodes an Evolutionarily Conserved Homeostasis, Mitosis, and Meiosis Regulator Rather than a Hyaluronan Receptor." *Cells*, 9 (4). https://doi.org/10.3390/cells9040819.

Howell, M. D. & Gottschall, P. E. (2012). "Lectican Proteoglycans, Their Cleaving Metalloproteinases, and Plasticity in the Central Nervous System Extracellular Microenvironment." *Neuroscience*, 217, 6–18. https://doi.org/10.1016/j.neuroscience.2012.05.034.

Humphrey, Jay D., Eric R. Dufresne. & Martin A. Schwartz. (2014). "Mechanotransduction and Extracellular Matrix Homeostasis." *Nature Reviews Molecular Cell Biology*, 15 (12), 802–12. https://doi.org/10.1038/nrm3896.

Jackson, David G. (2019). "Hyaluronan in the Lymphatics: The Key Role of the Hyaluronan Receptor LYVE-1 in Leucocyte Trafficking." *Matrix Biology*, 78–79, 219–35. https://doi.org/10.1016/j.matbio.2018.02.001.

Johnson, Pauline, Arif A. Arif, Sally S.M. Lee-Sayer. & Yifei Dong. (2018). "Hyaluronan and Its Interactions with Immune Cells in the Healthy and Inflamed Lung." *Frontiers in Immunology*. https://doi.org/10.3389/fimmu.2018.02787.

Johnsson, Hans, Lars Eriksson, Bengt Gerdin, Roger Hällgren. & Gunnar Sedin. (2003). "Hyaluronan in the Human Neonatal Lung: Association with Gestational Age and Other Perinatal Factors." *Biology of the Neonate*, 84, 194–201. https://doi.org/10.1159/000072302.

Johnsson, Hans, Lars Eriksson, Anders Jonzon, Torvard C. Laurent. & Gunnar Sedin. (1998). "Lung Hyaluronan and Water Content in Preterm and Term Rabbit Pups Exposed to Oxygen or Air." *Pediatric Research*, 44, 716–22. https://doi.org/10.1203/00006450-199811000-00014.

Kanayama, Naohiro, Junko Goto. & Toshihiko Terao. (1999). "The Role of Low Molecular Weight Hyaluronic Acid Contained in Wharton's Jelly in Necrotizing Funisitis." *Pediatric Research*, 45, 510–14. https://doi.org/10.1203/00006450-199904010-00009.

Kern, Christine B., Russell A. Norris, Robert P. Thompson, W. Scott Argraves, Sarah E. Fairey, Leticia Reyes, Stanley Hoffman, Roger R. Markwald. & Corey H. Mjaatvedt. (2007). "Versican Proteolysis Mediates Myocardial Regression during Outflow Tract Development." *Developmental Dynamics*, 236 (3), 671–83. https://doi.org/10.1002/dvdy.21059.

Kern, Christine B., Waleed O. Twal, Corey H. Mjaatvedt, Sarah E. Fairey, Bryan P. Toole, M. Luisa Iruela-Arispe. & Scott Argraves, W. (2006). "Proteolytic Cleavage of Versican during Cardiac Cushion Morphogenesis." *Developmental Dynamics*, 235 (8), 2238–47. https://doi.org/10.1002/dvdy.20838.

Kern, Christine B., Andy Wessels, Jessica McGarity, Laura J. Dixon, Ebony Alston, W. Scott Argraves, Danielle Geeting, Courtney M. Nelson, Donald R. Menick. & Suneel S. Apte. (2010). "Reduced Versican Cleavage Due to Adamts9 Haploinsufficiency Is Associated with Cardiac and Aortic Anomalies." *Matrix Biology*, 29 (4), 304–16. https://doi.org/10.1016/j.matbio.2010.01.005.

Kobayashi, Nobutaka, Seiji Miyoshi, Takahide Mikami, Hiroshi Koyama, Masato Kitazawa, Michiko Takeoka, Kenji Sano, et al. (2010). "Hyaluronan Deficiency in Tumor Stroma Impairs Macrophage Trafficking and Tumor Neovascularization." *Cancer Research*, 70 (18). https://doi.org/10.1158/0008-5472.CAN-09-4687.

Koch, Christopher D., Chan Mi Lee. & Suneel S. Apte. (2020). "Aggrecan in Cardiovascular Development and Disease." *Journal of Histochemistry and Cytochemistry*. https://doi.org/10.1369/0022155420952902.

Kurpios, Natasza A, Marta Iban, Nicole M Davis, Wei Lui, Tamar Katz. & James F Martin. (2008). "The Direction of Gut Looping Is Established by Changes in the Extracellular Matrix and in Cell : Cell Adhesion." *Proceedings of the National Academy of Sciences of the United States of America*, 105 (25), 8499–8506.

Landauer, Walter. (1965). "Nanomelia, a Lethal Mutation of the Fowl." *Journal of Heredity*, 56 (3). https://doi.org/10.1093/oxfordjournals.jhered.a107392.

Landolt, R. M., Vaughan, L., Winterhalter, K. H. & Zimmermann, D. R. (1995). "Versican Is Selectively Expressed in Embryonic Tissues That Act as Barriers to Neural Crest Cell Migration and Axon Outgrowth." *Development*, 121 (8).

Lauer, Mark E, Tibor T Glant, Katalin Mikecz, Paul L DeAngelis, F Michael Haller, M Elaine Husni, Vincent C Hascall. & Anthony Calabro. (2013). "Irreversible Heavy Chain Transfer to Hyaluronan Oligosaccharides by Tumor Necrosis Factor-Stimulated Gene-6." *The Journal of Biological Chemistry*, 288 (1), 205–14. https://doi.org/10.1074/jbc.M112.403998.

Leng, Yue, Ammara Abdullah, Michael K. Wendt. & Sarah Calve. (2019). "Hyaluronic Acid, CD44 and RHAMM Regulate Myoblast Behavior during Embryogenesis." *Matrix Biology*, 78–79, 236–54. https://doi.org/10.1016/j.matbio.2018.08.008.

Li, H., Schwartz, N. B. & Vertel, B. M. (1993). "CDNA Cloning of Chick Cartilage Chondroitin Sulfate (Aggrecan) Core Protein and Identification of a Stop Codon in the Aggrecan Gene Associated with the Chondrodystrophy, Nanomelia." *Journal of Biological Chemistry*, 268 (31). https://doi.org/10.1016/s0021-9258(19)49491-x.

Liang, Jiurong, Dianhua Jiang, Jason Griffith, Shuang Yu, Juan Fan, Xiaojian Zhao, Richard Bucala. & Paul W. Noble. (2007). "CD44 Is a Negative Regulator of Acute Pulmonary Inflammation and Lipopolysaccharide-TLR Signaling in Mouse Macrophages." *The Journal of Immunology*, 178 (4). https://doi.org/10.4049/jimmunol.178.4.2469.

Liang, Zhongxing, Joann Brooks, Margaret Willard, Ke Liang, Younghyoun Yoon, Seunghee Kang. & Hyunsuk Shim. (2007). "CXCR4/CXCL12 Axis Promotes VEGF-Mediated Tumor Angiogenesis through Akt Signaling Pathway." *Biochemical and Biophysical Research Communications*, 359 (3). https://doi.org/10.1016/j.bbrc.2007.05.182.

Lieleg, Oliver, & Katharina Ribbeck. (2011). "Biological Hydrogels as Selective Diffusion Barriers." *Trends in Cell Biology.* https://doi.org/10.1016/j.tcb.2011.06.002.

Logan, Malcolm, Sylvia M. Pagán-Westphal, Devyn M. Smith, Laura Paganessi. & Clifford J. Tabin. (1998). "The Transcription Factor Pitx2 Mediates Situs-Specific Morphogenesis in Response to Left-Right Asymmetric Signals." *Cell*, 94 (3). https://doi.org/10.1016/S0092-8674(00)81474-9.

Lu, Pengfei, Valerie M. Weaver. & Zena Werb. (2012). "The Extracellular Matrix: A Dynamic Niche in Cancer Progression." *Journal of Cell Biology*, 196 (4), 395–406. https://doi.org/10.1083/jcb.201102147.

Lumsden, A. G. S. (1979). "Pattern Formation in the Molar Dentition of the Mouse." *Journal de Biologie Buccale*, 7 (1).

Mack, Judith A., Ron J. Feldman, Naoki Itano, Koji Kimata, Mark Lauer, Vincent C. Hascall. & Edward V. Maytin. (2012). "Enhanced Inflammation and Accelerated Wound Closure Following Tetraphorbol Ester Application or Full-Thickness Wounding in Mice Lacking Hyaluronan Synthases Has1 and Has3." *Journal of Investigative Dermatology*, 132 (1). https://doi.org/10.1038/jid.2011.248.

Markasz, Laszlo, Rashmin C. Savani, Gunnar Sedin. & Richard Sindelar. (2018). "The Receptor for Hyaluronan-Mediated Motility (RHAMM) Expression in Neonatal Bronchiolar Epithelium Correlates Negatively with Lung Air Content." *Early Human Development*, 127, 58–68. https://doi.org/10.1016/j.earlhumdev.2018.10.002.

Matsumoto, Kazu, Yingcui Li, Caroline Jakuba, Yoshinori Sugiyama, Tetsuya Sayo, Misako Okuno, Caroline N. Dealy., et al. (2009). "Conditional Inactivation of Has2 Reveals a Crucial Role for Hyaluronan in Skeletal Growth, Patterning, Chondrocyte Maturation and Joint Formation in the Developing Limb." *Development*, 136 (16), 2825–35. https://doi.org/10.1242/dev.038505.

McCulloch, Daniel R., Courtney M. Nelson, Laura J. Dixon, Debra L. Silver, James D. Wylie, Volkhard Lindner, Takako Sasaki, Marion A. Cooley, W. Scott Argraves. & Suneel S. Apte. (2009). "ADAMTS Metalloproteases Generate Active Versican Fragments That

Regulate Interdigital Web Regression." *Developmental Cell*, 17 (5), 687–98. https://doi.org/10.1016/j.devcel.2009.09.008.

Milev, Peter, Patrice Maurel, Atsuro Chiba, Markus Mevissen, Susanna Popp, Yu Yamaguchi, Renée K. Margolis. & Richard U. Margolis. (1998). "Differential Regulation of Expression of Hyaluronan-Binding Proteoglycans in Developing Brain: Aggrecan, Versican, Neurocan, and Brevican." *Biochemical and Biophysical Research Communications*, 247, 207–12. https://doi.org/10.1006/bbrc.1998.8759.

Miranović, Vesna. (2014). "The Incidence of Congenital Heart Disease: Previous Findings and Perspectives." *Srpski Arhiv Za Celokupno Lekarstvo*, 142 (3–4). https://doi.org/10.2298/SARH1404243M.

Misra, Suniti, Vincent C Hascall, Roger R Markwald. & Shibnath Ghatak. (2015). "Interactions between Hyaluronan and Its Receptors (CD44, RHAMM) Regulate the Activities of Inflammation and Cancer." *Frontiers in Immunology*, 6, 201. https://doi.org/10.3389/fimmu.2015.00201.

Mittal, Nishant, Sung Han Yoon, Hirokazu Enomoto, Miyama Hiroshi, Atsushi Shimizu, Atsushi Kawakami, Misato Fujita, Hideto Watanabe, Keiichi Fukuda. & Shinji Makino. (2019). "Versican Is Crucial for the Initiation of Cardiovascular Lumen Development in Medaka (Oryzias Latipes)." *Scientific Reports*, 9, (9475).

Mjaatvedt, C. H., Yamamura, H., Capehart, A. A., Turner, D. & Markwald, R. R. (1998). "The Cspg2 Gene, Disrupted in the Hdf Mutant, Is Required for Right Cardiac Chamber and Endocardial Cushion Formation." *Developmental Biology*, 202 (1), 56–66. https://doi.org/10.1006/DBIO.1998.9001.

Mori, Mayumi, Agnes Bogdan, Timea Balassa, Timea Csabai. & Júlia Szekeres-Bartho. (2016). "The Decidua—the Maternal Bed Embracing the Embryo—Maintains the Pregnancy." *Seminars in Immunopathology*. https://doi.org/10.1007/s00281-016-0574-0.

Mueller, Andre Michael, Bo Hyung Yoon. & Saud Ahmed Sadiq. (2014). "Inhibition of Hyaluronan Synthesis Protects against Central Nervous System (CNS) Autoimmunity and Increases CXCL12 Expression in the Inflamed CNS." *Journal of Biological Chemistry*, 289 (33), 22888–99. https://doi.org/10.1074/jbc.M114.559583.

Muggenthaler, Martina M. A., Biswajit Chowdhury, S. Naimul Hasan, Harold E. Cross, Brian Mark, Gaurav V. Harlalka, Michael A. Patton, et al. (2017). "Mutations in HYAL2, Encoding Hyaluronidase 2, Cause a Syndrome of Orofacial Clefting and Cor Triatriatum Sinister in Humans and Mice." *PLoS Genetics*, 13 (1). https://doi.org/10.1371/journal.pgen.1006470.

Nandadasa, Sumeda, Anna O'Donnell, Ayako Murao, Yu Yamaguchi, Ronald J. Midura, Lorin Olson. & Suneel S. Apte. (2021). "The Versican-Hyaluronan Complex Provides an Essential Extracellular Matrix Niche for Flk1+ Hematoendothelial Progenitors: Versican and Hyaluronan in Vasculogenesis." *Matrix Biology*. https://doi.org/10.1016/j.matbio.2021.01.002.

Pardue, Erin L., Samir Ibrahim. & Anand Ramamurthi. (2008). "Role of Hyaluronan in Angiogenesis and Its Utility to Angiogenic Tissue Engineering." *Organogenesis*, 4 (4), 203–214. https://doi.org/10.4161/org.4.4.6926.

Peal, David S., Stacey N. Lynch. & David J. Milan. (2011). "Patterning and Development of the Atrioventricular Canal in Zebrafish." *Journal of Cardiovascular Translational Research*, 4 (6). https://doi.org/10.1007/s12265-011-9313-z.

Petrey, Aaron C. & Carol A. de la Motte. (2014). "Hyaluronan, a Crucial Regulator of Inflammation." *Frontiers in Immunology*, 5. https://doi.org/10.3389/fimmu.2014.00101.

Pratt, R. M., Larsen, M. A. & Johnston, M. C. (1975). "Migration of Cranial Neural Crest Cells in a Cell-Free Hyaluronate-Rich Matrix." *Developmental Biology*, 44 (2). https://doi.org/10.1016/0012-1606(75)90400-5.

Primorac, D., Stover, M. L., Clark, S. H. & Rowe, D. W. (1994). "Molecular Basis of Nanomelia, a Heritable Chondrodystrophy of Chicken." *Matrix Biology*, 14 (4). https://doi.org/10.1016/0945-053X(94)90195-3.

Protin, U., Schweighoffer, T., Jochum, W. & Hilberg, F. (1999). "CD44-Deficient Mice Develop Normally with Changes in Subpopulations and Recirculation of Lymphocyte Subsets." *Journal of Immunology (Baltimore, Md. : 1950)*, 163 (9).

Rao, Cherie, Dorothee Foernzler, Stacie K. Loftus, Shanming Liu, John D. McPherson, Katherine A. Jungers, Suneel S. Apte, William J.

Pavan. & David R. Beier. (2003). "A Defect in a Novel ADAMTS Family Member Is the Cause of the Belted White-Spotting Mutation." *Development*, 130 (19). https://doi.org/10.1242/dev.00668.

Rodgers, Laurel S., Sofia Lalani, Katharine M. Hardy, Xueyu Xiang, Derrick Broka, Parker B. Antin. & Todd D. Camenisch. (2006). "Depolymerized Hyaluronan Induces Vascular Endothelial Growth Factor, a Negative Regulator of Developmental Epithelial-to-Mesenchymal Transformation." *Circulation Research*, 99 (6), 583–89. https://doi.org/10.1161/01.RES.0000242561.95978.43.

Rozario, Tania. & Douglas W. DeSimone. (2010). "The Extracellular Matrix in Development and Morphogenesis: A Dynamic View." *Developmental Biology*, 341, 126–40. https://doi.org/10.1016/j.ydbio.2009.10.026.

Russell, Darryl L., Hannah M. Brown. & Kylie R. Dunning. (2015). "ADAMTS Proteases in Fertility." *Matrix Biology*, 44–46, (May), 54–63. https://doi.org/10.1016/J.MATBIO.2015.03.007.

Ryan, Aimee K., Bruce Blumberg, Concepción Rodriguez-Esteban, Sayuri Yonei-Tamura, Koji Tamura, Tohru Tsukui, Jennifer De La Peña, et al. (1998). "Pitx2 Determines Left-Right Asymmetry of Internal Organs in Vertebrates." *Nature*, 394 (6693). https://doi.org/10.1038/29004.

Ryckebüsch, Lucile, Lydia Hernandez, Carole Wang, Jenny Phan. & Deborah Yelon. (2016). "Tmem2 Regulates Cell-Matrix Interactions That Are Essential for Muscle Fiber Attachment." *Development*, 143, 2965–72.

Salustri, Antonietta, Masaki Yanagishita, Charles B. Underhill, Torvard C. Laurent. & Vincent C. Hascall. (1992). "Localization and Synthesis of Hyaluronic Acid in the Cumulus Cells and Mural Granulosa Cells of the Preovulatory Follicle." *Developmental Biology*, 151 (2), 541–51. https://doi.org/10.1016/0012-1606(92)90192-J.

Sánchez, Natalia, María Constanza González-Ramírez, Esteban G. Contreras, Angélica Ubilla, Jingjing Li, Anyeli Valencia, Andrés Wilson, Jeremy B.A. Green, Abigail S. Tucker. & Marcia Gaete. (2020). "Balance Between Tooth Size and Tooth Number Is

Controlled by Hyaluronan." *Frontiers in Physiology*, 11. https://doi.org/10.3389/fphys.2020.00996.

Savin, Thierry, Natasza A. Kurpios, Amy E. Shyer, Patricia Florescu, Haiyi Liang, L. Mahadevan. & Clifford J. Tabin. (2011). "On the Growth and Form of the Gut." *Nature*, 476, 57–62. https://doi.org/10.1038/nature10277.

Shin, Masahiro, Timothy J. Beane, Aurelie Quillien, Ira Male, Lihua J. Zhu. & Nathan D. Lawson. (2016). "Vegfa Signals through ERK to Promote Angiogenesis, but Not Artery Differentiation." *Development (Cambridge)*, 143 (20). https://doi.org/10.1242/dev.137919.

Shin, Masahiro, Ira Male, Timothy J. Beane, Jacques A. Villefranc, Fatma O. Kok, Lihua J. Zhu. & Nathan D. Lawson. (2016). "Vegfc Acts through ERK to Induce Sprouting and Differentiation of Trunk Lymphatic Progenitors." *Development (Cambridge)*, 143 (20). https://doi.org/10.1242/dev.137901.

Silva, Ana Catarina, Cassilda Pereira, Ana Catarina R. G. Fonseca, Perpétua Pinto-do-Ó. & Diana S. Nascimento. (2021). "Bearing My Heart: The Role of Extracellular Matrix on Cardiac Development, Homeostasis, and Injury Response." *Frontiers in Cell and Developmental Biology*. https://doi.org/10.3389/fcell.2020.621644.

Silver, Debra L., Ling Hou, Robert Somerville, Mary E. Young, Suneel S. Apte. & William J. Pavan. (2008). "The Secreted Metalloprotease ADAMTS20 Is Required for Melanoblast Survival." *PLoS Genetics*, 4. https://doi.org/10.1371/journal.pgen.1000003.

Sivakumar, Aravind, Aparna Mahadevan, Mark E. Lauer, Ricky J. Narvaez, Siddesh Ramesh, Cora M. Demler, Nathan R. Souchet, et al. (2018). "Midgut Laterality Is Driven by Hyaluronan on the Right." *Developmental Cell*, 46 (5), 533-551.e5. https://doi.org/10.1016/j.devcel.2018.08.002.

Slevin, Mark, Jurek Krupinski, John Gaffney, Sabine Matou, David West, Horace Delisser, Rashmin C. Savani. & Shant Kumar. (2007). "Hyaluronan-Mediated Angiogenesis in Vascular Disease: Uncovering RHAMM and CD44 Receptor Signaling Pathways." *Matrix Biology*, 26 (1), 58–68. https://doi.org/10.1016/j.matbio.2006.08.261.

Smith, Kelly A., Anne K. Lagendijk, Andrew D. Courtney, Huijun Chen, Scott Paterson, Benjamin M. Hogan, Carol Wicking. & Jeroen Bakkers. (2011). "Transmembrane Protein 2 (Tmem2) Is Required to Regionally Restrict Atrioventricular Canal Boundary and Endocardial Cushion Development." *Development*, 138, 4193–98. https://doi.org/10.1242/dev.065375.

Smith, Lucas R., Sangkyun Cho. & Dennis E. Discher. (2018). "Stem Cell Differentiation Is Regulated by Extracellular Matrix Mechanics." *Physiology*, 33 (1), 16–25. https://doi.org/10.1152/physiol.00026.2017.

Stankunas, Kryn, Calvin T. Hang, Zhi Yang Tsun, Hanying Chen, Nathan V. Lee, Jiang I. Wu, Ching Shang, et al. (2008). "Endocardial Brg1 Represses ADAMTS1 to Maintain the Microenvironment for Myocardial Morphogenesis." *Developmental Cell*, 14 (2), 298–311. https://doi.org/10.1016/j.devcel.2007.11.018.

Totong, Ronald, Thomas Schell, Fabienne Lescroart, Lucile Ryckebüsch, Yi Fan Lin, Tomasz Zygmunt, Lukas Herwig, et al. (2011). "The Novel Transmembrane Protein Tmem2 Is Essential for Coordination of Myocardial and Endocardial Morphogenesis." *Development*, 138, 4199–4205. https://doi.org/10.1242/dev.064261.

Turley, Eva A., Paul W. Noble. & Lilly Y. W. Bourguignon. (2002). "Signaling Properties of Hyaluronan Receptors." *Journal of Biological Chemistry*, 277 (7), 4589–92. https://doi.org/10.1074/jbc.R100038200.

Underhill, Charles B., Huong A. Nguyen, Mehran Shizari. & Martine Culty. (1993). "CD44 Positive Macrophages Take up Hyaluronan during Lung Development." *Developmental Biology*, 155 (2). https://doi.org/10.1006/dbio.1993.1032.

Vertel, B. M., Grier, B. L., Li, H. & Schwartz, N. B. (1994). "The Chondrodystrophy, Nanomelia: Biosynthesis and Processing of the Defective Aggrecan Precursor." *Biochemical Journal* 301 (1). https://doi.org/10.1042/bj3010211.

Wang, Xia, Chunman Li, Zeyao Zhu, Li Yuan, Wood Yee Chan. & Ou Sha. (2020). "Extracellular Matrix Remodeling During Palate Development." *Organogenesis*. https://doi.org/10.1080/15476278.2020.1735239.

Watanabe, Hideto, Koji Kimata, Sergio Line, Dave Strong, Luo Yi Gao, Christine A. Kozak. & Yoshihiko Yamada. (1994). "Mouse Cartilage Matrix Deficiency (Cmd) Caused by a 7 Bp Deletion in the Aggrecan Gene." *Nature Genetics*, 7, 154–57. https://doi.org/10.1038/ng0694-154.

Watanabe, Hideto, Ken Nakata, Koji Kimata, Isao Nakanishi. & Yoshihiko Yamada. (1997). "Dwarfism and Age-Associated Spinal Degeneration of Heterozygote Cmd Mice Defective in Aggrecan." *Proceedings of the National Academy of Sciences of the United States of America*, 94 (13). https://doi.org/10.1073/pnas.94.13.6943.

Watanabe, Hideto. & Yoshihiko Yamada. (2002). "Chondrodysplasia of Gene Knockout Mice for Aggrecan and Link Protein." *Glycoconjugate Journal*. https://doi.org/10.1023/A:1025344332099.

Welsh, Ian C., Michael Thomsen, David W. Gludish, Catalina Alfonso-Parra, Yan Bai, James F. Martin. & Natasza A. Kurpios. (2013). "Integration of Left-Right Pitx2 Transcription and Wnt Signaling Drives Asymmetric Gut Morphogenesis via Daam2." *Developmental Cell*, 26 (6), 629–44. https://doi.org/10.1016/j.devcel.2013.07.019.

Wolf, Zena T., Harrison A. Brand, John R. Shaffer, Elizabeth J. Leslie, Boaz Arzi, Cali E. Willet, Timothy C. Cox., et al. (2015). "Genome-Wide Association Studies in Dogs and Humans Identify ADAMTS20 as a Risk Variant for Cleft Lip and Palate." *PLoS Genetics*, 11 (3). https://doi.org/10.1371/journal.pgen.1005059.

Yamaguchi, Yu, Hayato Yamamoto, Yuki Tobisawa. & Fumitoshi Irie. (2019). "TMEM2: A Missing Link in Hyaluronan Catabolism Identified?" *Matrix Biology*. https://doi.org/10.1016/j.matbio.2018.03.020.

Yamamoto, Hayato, Yuki Tobisawa, Toshihiro Inubushi, Fumitoshi Irie, Chikara Ohyama. & Yu Yamaguchi. (2017). "A Mammalian Homolog of the Zebrafish Transmembrane Protein 2 (TMEM2) Is the Long-Sought-after Cell-Surface Hyaluronidase." *The Journal of Biological Chemistry*, 292 (18), 7304–13. https://doi.org/10.1074/jbc.M116.770149.

Yamamura, H., Zhang, M., Markwald, R. R. & Mjaatvedt, C. H. (1997). "A Heart Segmental Defect in the Anterior-Posterior Axis of a

Transgenic Mutant Mouse." *Developmental Biology*, 186 (1), 58–72. https://doi.org/10.1006/DBIO.1997.8559.

Yonemitsu, Marisa A., Tzu yin Lin. & Kai Yu. (2020). "Hyaluronic Acid Is Required for Palatal Shelf Movement and Its Interaction with the Tongue during Palatal Shelf Elevation." *Developmental Biology*, 457 (1). https://doi.org/10.1016/j.ydbio.2019.09.004.

Yoshida, H., Nagaoka, A., Kusaka-Kikushima, A., M. Tobiishi, A., Kawabata, K., Sayo, T., Sakai, S., et al. (2013). "KIAA1199, a Deafness Gene of Unknown Function, Is a New Hyaluronan Binding Protein Involved in Hyaluronan Depolymerization." *Proceedings of the National Academy of Sciences*, 110 (14), 5612–17. https://doi.org/10.1073/pnas.1215432110.

In: Hyaluronic Acid
Editor: Vittorio Unfer

ISBN: 978-1-53619-743-3
© 2021 Nova Science Publishers, Inc.

Chapter 7

HYALURONIC ACID IN THE DEVELOPMENT OF THE GUT AND PROTECTION AGAINST NECROTIZING ENTEROCOLITIS

Kathryn Y. Burge, Jeffrey V. Eckert and Hala Chaaban[*]

Department of Pediatrics, Neonatal-Perinatal Medicine,
The University of Oklahoma Health Sciences Center

ABSTRACT

Necrotizing enterocolitis, a life-threatening intestinal inflammatory disorder primarily affecting premature infants, is a significant cause of morbidity and mortality in the postnatal period. Physiological factors contributing to increased risk for NEC include developmental immaturity of the mucosal immune system coinciding with rapid cell differentiation in the fetal intestine. Increasing evidence suggests HA enhances the development of the fetal and postnatal gut, as well as protects against severe gastrointestinal illnesses, such as NEC. When internalized by enterocytes or immune cells, HMW-HA is digested into O-HA or LMW-HA and transported through the blood within lysosomes. On

---

[*] Corresponding Author's Email: hala-chaaban@ouhsc.edu.

the contrary, when absorbed by microfold cells (M cells), HMW-HA enters the lymphatic system unchanged and is then dispersed systemically.

1. INTRODUCTION

Hyaluronic acid (HA) is a natural glycosaminoglycan (GAG) composed of linear, repeating residues of D-glucuronic acid and *N*-acetyl-D-glucosamine joined by glycosidic bonds (Kogan et al. 2007). HA is unique among GAGs in that it is non-sulfated, can reach a very high molecular weight (MW), is not bound to a core protein, and is not, therefore, synthesized by the Golgi complex (Kogan et al. 2007). HA is found ubiquitously in the human body and is a primary component of the extracellular matrix (ECM), where it aids tissue in retaining water (Papakonstantinou, Roth, and Karakiulakis 2012) through its negatively charged salt complexes, often referred to as hyaluronan or hyaluronate (Laurent 1989). High molecular weight (HMW) HA (>800 kDa) is produced endogenously by most vertebrate cell types, often disproportionately by mesenchymal cells, through HA synthases (HAS) located on the inner plasma membrane (Prehm 1990). Due to its large size, as the HA polymer is synthesized, it is simultaneously extruded through pores in the plasma membrane and deposited within pericellular jelly or ECM (Volpi et al. 2009). Once released, HA may be fragmented, internalized locally through cell surface receptor binding, or removed via lymph (Monslow, Govindaraju, and Puré 2015). The turnover of HA in the body is rapid (Volpi et al. 2009), on the order of days in most tissues, with depolymerization of HA accomplished primarily via tissue-specific hyaluronidases (HYAL), cell migration-inducing and hyaluronan-binding protein (CEMIP), transmembrane protein 2 (TMEM2), or through non-specific oxidative degradation by reactive oxygen species (ROS) (Papakonstantinou, Roth, and Karakiulakis 2012). The latter depolymerization via ROS may, in part, explain much of the antioxidant activity attributed to HA (Fallacara et al. 2018). Importantly, the ratio of endogenous HMW-HA synthesis to degradation through fragmentation is often associated with, or can in part determine, disease progression (Barbosa de Souza et al. 2020), with the degradation rate often accounting for more substantial variability within the ratio (Monslow, Govindaraju, and Puré 2015).

The biological roles of HA, whether produced endogenously, subsequently fragmented, or introduced exogenously, appear to depend upon the length, and by extension, the MW, of the polymer (Stern, Asari, and Sugahara 2006). HA polymer MW ranges have not been standardized by the field and are often defined arbitrarily, often a source of complication (Cyphert, Trempus, and Garantziotis 2015). In general, HMW-HA synthesized by HASs enhances tissue hydration, serves as an antioxidant, and inhibits endothelial cell growth through physical limitation of ECM space, while oligo-HA (<10 kDa), low MW (LMW) HA (10-300 kDa), and medium MW (MMW) HA (300-800 kDa), often produced via endogenous fragmentation of larger HA polymers following tissue injury, induce wound healing processes, activate the innate immune system, produce growth factors, and stimulate angiogenesis (Riehl, Ee, and Stenson 2012, Fallacara et al. 2018, Monslow, Govindaraju, and Puré 2015, Papakonstantinou, Roth, and Karakiulakis 2012, Slevin et al. 2007, Jackson 2009, Jiang, Liang, and Noble 2011, Soltés et al. 2006). Thus, in addition to playing a role in inflammatory and fibrotic diseases, these lower MW HA polymers also appear to aid wound healing and tissue repair (Slevin, Kumar, and Gaffney 2002). HMW-HA may also encourage wound healing by providing soft scaffolding immediately following tissue injury, upon which lower MW-HA fragments may increase cell migration, angiogenesis, and tissue remodeling (Monslow, Govindaraju, and Puré 2015). In addition, HMW-HA may promote HA receptor CD44 (cluster of differentiation-44) crosslinking, inducing synthesis of anti-inflammatory cytokines and spurring resolution of wound inflammation (Ruppert et al. 2014). Cell signaling attributed to HA fragments, independent of fragment size, is often considered to be contingent upon binding to HA receptors, such as CD44, layilin, TLR2 (Toll-like receptor 2), and TLR4, several of which are widely expressed throughout the gastrointestinal tract (Riehl, Ee, and Stenson 2012). However, HMW-HA has also been theorized to influence signal transduction through physically limiting access to the cell surface through its accumulation in the ECM or pericellular jelly (Ebid, Lichtnekert, and Anders 2014), and thus internalization of the polymer may not be required to exert physiological effects (Kim and de la Motte 2020). This phenomenon has been demonstrated *in vitro*, where HMW-HA competitively inhibited LMW-HA TLR2 signaling (Scheibner et al. 2006). Studies of the effects of HA receptor-binding *in vivo* have been complicated by a lack of accurate tissue-specific mapping of HA receptor expression, ontogenetically or at developmental maturity, throughout the intestinal epithelium, confounding any conclusions relating to size-specific differences in HA receptor-binding (Kim et al. 2018).

The degree to which oral HA fragment supplementation can be absorbed in the intestine has long been debated. Kimura et al. demonstrated oral HA 300 kDa and 2000 kDa was fully absorbed in the colon, even if the responsible mechanism was not yet clarified. In several studies, the distribution of HA to the skin and serum was shown (Kimura et al. 2016). Oe et al. orally administered HA 920 kDa to rats, where upwards of 90% was absorbed in the intestine, and subsequently metabolized by the skin (Oe et al. 2014). Balogh et al. confirmed orally administered HA (1000 kDa) is absorbed in rats and dogs using 99Technetium (99Tc) labeling, and subsequently distributed through the blood to the skin and joints (Balogh et al. 2008). Several authors have reported the ability of O-HA to permeate the intestinal barrier, passing passively between the enterocytes to the circulatory system, but this process is not likely applicable for HA with a molecular weight >10 kDa (de Souza, Chaud, and Santana 2019). While LMW-HA and MMW-HA absorption can be explained by interaction of these molecules with enterocyte membrane TLR4 and by subsequent lysosomal degradation, for HMW-HA a different mechanism has been hypothesized. The absorption of undegraded HMW-HA can occur due to the presence of microfold cells (M cells) in the intestinal epithelium, which transport the molecule to gut-associated lymphatic tissue (GALT) (de Souza, Chaud, and Santana 2019). The proposed molecular dynamics of HA intestinal uptake begins with interaction of HA with TLR4, where HA is internalized. Although several cells absorb HMW-HA after ingestion (enterocytes (Neal et al. 2006), M cells (Barthe, Woodley, and Houin 1999, Neutra 1998), dendritic cells and macrophages (Oe et al. 2016, Corr, Gahan, and Hill 2008)), only M cells can deliver intact HMW-HA to GALT (Rubas and Grass 1991).

The destination of orally ingested HMW-HA also differs by the cell type by which it is absorbed. When internalized by enterocytes or immune cells, HMW-HA is digested to O-HA or LMW-HA and transported via lysosomes to the blood. On the contrary, when absorbed by M cells, HMW-HA enters the lymphatic system unaltered, where it is distributed throughout the body (de Souza, Chaud, and Santana 2019). The mechanisms through which HA of varying sizes is absorbed and distributed throughout the body may influence the ways in which the postnatal intestine develops and the capability of the developing intestine, particularly that of premature infants, to defend against microbial infection and inflammation, as occurs with necrotizing enterocolitis (NEC).

2. HYALURONIC ACID IN HUMAN MILK

Human milk (HM) is a critically important source of exogenous HA during infancy. Both postnatal gastrointestinal development and protection against neonatal intestinal pathogens are likely hastened via HA supplied through HM, which reaches the small intestine intact given the lack of degradative enzyme activity in the proximal intestine (Maccari et al. 2016). Infants fed exclusively HM are subject to 50% fewer enteric infections compared with those fed formula (Grulee, Sanford, and Herron 1934, Howie et al. 1990). In preterm infants, specifically, even partial bovine-based feedings are associated with an increase in morbidity and mortality (Abrams et al. 2014), while very low birthweight (VLBW) infants experience a 6-fold decrease in the incidence of NEC when a minimum of 50% of feedings are derived from HM within 2 weeks postpartum (Sisk et al. 2007). Quantitative and qualitative differences in the HA composition of HM compared with bovine-based formula exist, and these differences may account for some of the increased risk of NEC development among formula-fed infants. In term HM, GAGs, in total, are nearly seven times more prevalent compared with bovine milk, the basis for most infant formula, and HA, in particular, is found at substantially higher levels (5.4 mg/L compared with 2.7 mg/L) (Coppa et al. 2011). In preterm HM, GAG levels are three times higher compared with term HM, but both total

GAG and HA content in term and preterm milk continually decline from a peak in colostrum (Coppa et al. 2013) out to, at minimum, the 6th month postpartum (Wang et al. 2018). Likewise, Hill et al. have reported a peak HA concentration of 755 ng/ml during the first week postpartum, an average HA concentration during the first 60 days postpartum of 452 ng/ml, and a relatively static HA concentration of 215 ng/ml for the remainder of the year (Hill et al. 2013). This peak of HA content in colostrum is potentially indicative of its requirement for proper neonatal development and maturation of the gastrointestinal tract. Yuan et al. determined the MW distribution of HA within HM, concluding HA over 110 kDa comprises 95% of the HA within HM, with LMW-HA accounting for the remaining 5% (Yuan et al. 2015). Interestingly, all HA extracted from HM was less than 1500 kDa, where HA extracted from most biological tissues to date, while polydisperse, has been of a much higher average MW above 2000 kDa (Armstrong and Bell 2002). The relatively low variability among HA size ranges among subjects seems to indicate a highly regulated balance between HA degradation and HA synthesis in the mammary gland (Yuan et al. 2015). Finally, differences in bioavailability of HA within HM and infant formula are evident, as the percentage of HA recovered in the feces of breastfed infants is nearly seven times lower than that of formula-fed infants (Maccari et al. 2016).

3. HYALURONIC ACID IN INTESTINAL DEVELOPMENT

The intestine must not only provide a means through which calories and nutrients are digested and absorbed, but also must regulate the degree of, and response to, interaction between the epithelium and commensal and pathogenic bacteria populating the intestinal lumen. To maintain this critical barrier function, the intestinal epithelium houses tight junctions, which selectively control the flow of luminal contents into the mucosa (Turner 2009). Additionally, a mucus lining secreted by epithelial goblet cells physically inhibits microbial adherence to enterocytes (Turner 2009), while the release of antimicrobial peptides (AMPs) from Paneth cells chemically deters microbes from binding to the epithelium (Clevers and Bevins 2013). Embedded within the intestine are immune cells, which depend upon early stimulation by luminal bacteria for proper development as the final layer of intestinal defense (Round and Mazmanian 2009). To best accomplish these many functions, the epithelium of the intestine is organized in finger-like luminal projections termed villi. Intestinal stem cells (ISCs) at the base of the villi (crypts) proliferate extensively to fully regenerate the epithelium within days, with new cells pushed toward villi tips while simultaneously terminally differentiating (Shyer et al. 2013). Once at the villi tips, differentiated cells undergo apoptosis and are released into the intestinal lumen (Chin et al. 2017). The mechanistic underpinnings of gastrointestinal tract proliferation and maturation, particularly postnatally, are poorly understood. While there have been a number of studies examining the ISC proliferative response to individual growth factors (e.g., Abo et al. 2020, Li et al. 2018), these experiments often ignore growth through crypt fission, whereby a founder crypt, through unspecified regulatory mechanisms, divides to produce daughter crypts.

Following fertilization, the relatively flat gut tissue transitions to a tube. The anterior-posterior axis is established through the release of morphogens, initiating organ-specific molecular expression in the foregut, midgut, and hindgut (de Santa Barbara, van den Brink, and Roberts 2003). Differential Wnt signaling along the length of the tube establishes gradients along which the duodenum, jejunum, ileum, and colon are distinguished (Chin et al. 2017). Between weeks 3 and 7 in human fetal intestinal development, the endoderm-derived gut tube, a pseudostratified epithelium at this stage (Grosse et al. 2011), begins to elongate and widen through generalized proliferation, smooth muscle cells differentiate and surround the tube (McHugh 1996), vagal neural crest cells migrate to the tube to innervate the tissue (Hao et al. 2016), and vascular smooth muscle cells differentiate from the mesothelium (Hatch and Mukouyama 2015). Around gestational week 8, the midgut rotates to create left-right orientation, and epithelial surface area is rapidly increased via the process of villus morphogenesis. Clusters of mesenchymal cells aggregate along the basal surface of the epithelium, pushing cells outward toward the lumen to create villi, with highly proliferative intervillous domains (analogous to adult crypts) between villi (Chin et al. 2017). The initiation of villus morphogenesis coincides with the appearance of the first leucine-rich repeat-containing G-protein receptor-positive (Lgr5$^+$) stem cells, phenotypic precursors to adult intestinal stem cells (Sprangers, Zaalberg, and Maurice 2021). Following epithelial organization into villi, intervillous domain proliferation is driven largely by Wnt/β-catenin signaling, while the subsequent differentiation of epithelial cells into secretory and absorptive enterocytes is differentiated via Notch signaling (Chin et al. 2017). Paneth cells, highly differentiated secretory enterocytes, are not distinguished within the human intestine until gestational week 20 (Mallow et al. 1996), while Peyer's patches, lymphocyte hubs, are established at 19 weeks (Sprangers, Zaalberg, and Maurice 2021). At birth, the neonatal intestine is structurally intact, but resident mucosal immune cells require colonization by the microbiome for the organ to become functionally mature. Postnatal intestinal lengthening occurs by a combination of crypt hyperplasia and crypt fission, with peak rates in human infants between 6 and 12 months postnatally (Cummins et al. 2008).

Hyaluronic acid is essential throughout embryogenesis and fetal development. Developing embryos are enveloped by matrices rich in HMW-HA. Counter to the effects of HMW-HA in most adult tissues, these matrices in the embryo allow for proper tissue growth via cellular proliferation and adhesion (Monslow, Govindaraju, and Puré 2015). While HAS1$^{-/-}$ and HAS3$^{-/-}$ mice are both viable and reproductively capable, HAS2$^{-/-}$ mice are embryonically lethal at age E9.5 (approximately human gestational week 3). These mice deficient in HAS2 lack 96% of the HA of their wild-type counterparts, with most of the deficits occurring within the ECM (Camenisch et al. 2000). HA is also critical to left-right intestinal asymmetry, the basis for proper midgut rotation. Accumulation of HA within the right ECM of the gut tube initiates left-right asymmetry through a tumor necrosis factor-α-stimulated gene 6 (Tsg6)-dependent mechanism, allowing for leftward gut tilt and vascular exclusion on the right side (Sivakumar et al. 2018). Within the right ECM, Tsg6 expression modifies HA with heavy-chain peptides, creating an extremely stable matrix. Of importance, while HA is synthesized in much smaller quantities on the left side of the gut tube, HA on the left is required in driving angiogenesis, where HA accumulation on the right is antiangiogenic.

The importance of HA throughout fetal development is evidenced through levels provided through amniotic fluid. Interestingly, the highest levels of HA within the amniotic fluid are seen during 16-20 weeks of gestation (Dahl, Dahl, and Børresen 1986), coinciding with rapid cell differentiation in the fetal intestine. In addition, HA at 16 weeks is predominantly of medium and low MW, and without degradative enzymes in amniotic fluid, these are likely the sizes at which HA is synthesized (Dahl, Dahl, and Børresen 1986). Postnatally, HA signaling through CD44 and TLR4 appears to be a unique requirement of gastrointestinal development, as $CD44^{-/-}$ and $TLR4^{-/-}$ mice grow normally and do not experience weight loss despite impaired intestinal elongation (Riehl et al. 2015). During intestinal crypt development and growth associated with epithelial maintenance and renewal, extracellular HA is generally confined to a relatively thin layer along the epithelial basolateral surface of the crypts (Zheng, Riehl, and Stenson 2009). Rodent intestinal development models have demonstrated exogenous HMW-HA is deposited sequentially up the crypt-villus axis (Riehl, Ee, and Stenson 2012).

Riehl et al. examined intestinal development of weanling mouse pups following 5 weeks of treatment with either intraperitoneal HA 750 kDa or PEP-1, an HA-blocking peptide (Riehl, Ee, and Stenson 2012). Interestingly, the beginning of the treatment period coincided with the developmental period of intestinal crypt fission giving way to increased intestinal length (Dehmer et al. 2011). Endogenous HA blockade by PEP-1 significantly reduced intestinal length and growth factor expression, while exogenous HA administration increased growth factor expression with no resultant increase in intestinal length. HA administration increased jejunal crypt depth and villus height, as well as colonic crypt depth, via epithelial enterocyte proliferation, while PEP-1 administration had the opposite effect. Finally, PEP-1 decreased the gross number of crypts and villi, while exogenous HA administration did not alter crypt or villus numbers. Of note, HA 750 kDa supplementation also spurred induction of both HYAL3 and HAS1, presumably both increasing HA synthesis of both LMW fragments and HMW deposition, respectively. This study demonstrates the requirement for HA receptor-binding in normal intestinal development.

In further studies, Riehl et al. demonstrated small intestinal and colonic epithelial proliferation and apoptosis is dependent upon endogenous HA receptor binding of both CD44 and TLR4 *in vivo* (Riehl et al. 2015). Blockade of endogenous HA receptor binding reduced Lgr5$^+$ stem cell expansion in the crypts, decreased Paneth cell numbers, and led to a decline in crypt fission early in postnatal life. However, neither HA administration nor blockade of HA receptor binding in small intestinal enteroids *in vitro* recapitulated the *in vivo* effects of epithelial proliferation, potentially indicating Lgr5$^+$ stem cell expansion occurred through signaling outside the epithelium (Riehl et al. 2015). Recent work also demonstrates the fundamental role for HA in sustaining myenteric neuron homeostasis in rat colon, by forming a condensed matrix that surrounds the cells, similar to that found in the central nervous system (CNS) (Bistoletti et al. 2020, Karamanos et al. 2018). Accordingly, disruption of HA synthesis or deposition by pathological processes, intestinal inflammation or ischemia/reperfusion injury, for example (Filpa et al. 2017), negatively influences the efficiency of neurochemical coding, excitatory and inhibitory, resulting in the development of gastrointestinal dysmotility (Bistoletti et al. 2020). Altogether, the effects of hyaluronic acid on intestinal growth, development, and function are indispensable.

4. Hyaluronic Acid in Protection from Necrotizing Enterocolitis

Necrotizing enterocolitis, a life-threatening intestinal inflammatory disorder primarily affecting premature infants, is a significant cause of morbidity and mortality in the postnatal period. Rampant intestinal inflammation triggers abdominal distension and necrosis. Subsequent intestinal perforation can result in multi-organ failure and death (Neu and Walker 2011, Tanner et al. 2015). The etiology of the disease is complex, incorporating multiple risk factors beyond prematurity, such as the utilization of antibiotics (Dimmitt et al. 2010), dysbiosis (Pammi et al. 2017), or small for gestational age status. An additional risk factor, formula feeding, has been shown to increase both the incidence and severity of NEC (Samuels et al. 2017), and is particularly pertinent in the discussion of protection against NEC as HA is found at much higher levels in HM compared with infant formula (Coppa et al. 2011).

NEC pathogenesis is both highly complex and incompletely understood. Physiological factors contributing to increased risk for NEC include developmental immaturity of the mucosal immune system (Battersby and Gibbons 2013), altered microbiome (Pammi et al. 2017, Denning and Prince 2018), and a malfunctioning intestinal barrier (Moore et al. 2016). Excessive Gram-negative stimulation of TLR4 likely occurs in the preterm ileum (Krappmann et al. 2004), resulting in inflammatory cytokine and chemokine release, loss of enterocytes via apoptosis, inappropriate autophagy leading to a lack of replacement of these enterocytes, and excessive neutrophil recruitment, resulting in the functional breakdown of the intestinal barrier (Burge et al. 2019, Mara, Good, and Weitkamp 2018, Lu, Sodhi, and Hackam 2014, De Plaen et al. 2007, Markel et al. 2006). A leaky epithelium allows for increased bacterial translocation (Udall et al. 1981), inducing further inflammation through immune cell contact with bacterial antigens (Managlia et al. 2019) and encouraging the prototypical runaway inflammation of NEC (De Plaen 2013). The microvasculature, simultaneously stimulated through TLR4 activation, is subjected to reduced endothelial nitric oxide synthase (eNOS), contributing to NEC ischemia and necrosis (Good et al. 2016, Yazji et al. 2013, Watkins and Besner 2013). This excessive inflammation spreads systemically, negatively influencing even the brain (Thoma 2019).

The incorporation of GAGs, such as HA, into the pericellular and extracellular matrices of epithelial cells is widely considered to be essential in the functioning of the neonatal intestinal barrier (Bode et al. 2008). Interestingly, skip lesions of necrotic NEC tissue are associated with an impaired distribution of GAGs in the intestinal epithelium (Ade-Ajayi et al. 1996). Hyaluronic acid may positively affect many of the physiological processes underlying the pathogenesis of NEC. Glycosaminoglycans are known to interact with luminal bacterial binding of host cells, and HA is no exception. In fact, the bacteriostatic and bactericidal properties of HA are largely responsible for its widespread use in bioengineered orthopedic scaffolding (Carlson et al. 2004, Pirnazar et al. 1999). Coppa et al. treated both Caco-2 and Int-

407 cell lines with HM-derived GAGs and established a significant reduction in pathogenic bacterial adhesion, often the initial step in infection, to these enterocytes (Coppa et al. 2016). Additionally, Hill et al. (Hill et al. 2013) demonstrated reduced adhesion and invasion of *Salmonella typhimurium* in HT-29 intestinal epithelial monolayers following treatment by milk HA extract compared with either media alone or hyaluronidase-digested milk HA extract. Importantly, HA-containing media was removed prior to *Salmonella* introduction, implicating HA-induced changes to intestinal epithelial cells rather than a direct effect of HA on bacteria. The bacteriostatic and bactericidal effects of HA are noteworthy given the increasingly recognized role of a dysregulated microbiome on the pathogenesis of NEC. Similar to other human milk glycans (Koropatkin, Cameron, and Martens 2012), HA is not digested by human enzymes, suggesting that act as natural prebiotics to guide the development of the infant gut microbiota. In fact, Lee et al. demonstrated that HA bound to bilirubin leads to modulation of the composition of the microbiota of mice with dextran sodium sulfate (DSS) colitis and was associated with significantly improved bacterial richness, diversity, and a relative abundance of the protective bacteria *Clostridium* XIVα, known to induce regulatory T cells, and *Lactobacillus*, known to be protective in animal models of gut inflammation and NEC (Lee et al. 2020). Importantly, when mice were pretreated with broad-spectrum antibiotics, the protective effects of HA were significantly reduced, suggesting that the beneficial effects are at least, in part, attributed to the changes in the microbiome. Hyaluronic acid is also capable of upregulating intestinal defensive secretions of the host. Interestingly, treatment with low to medium MW HA appears to have a uniquely protective, anti-inflammatory effect on the intestinal epithelium not recapitulated in many other tissues (e.g., (Huang et al. 2010)). For example, Hill et al. noted increased expression of human β-defensin 2 through intestinal epithelium TLR4 signaling *in vitro* following administration of HA 35 kDa (Hill et al. 2012). β-defensin 2, an inducible, cationic, AMP produced by the intestinal epithelium, is characterized by broad antimicrobial activity, including against *Salmonella* and *Escherichia coli* (Baricelli et al. 2015). Induction of this AMP was not replicated by HA of lower or higher MW, however. In addition, when polydisperse HA was extracted directly from HM and

subsequently used to treat intestinal epithelial cells at physiological concentrations, a similar induction of human β-defensin 2 was demonstrated (Hill et al. 2013). Finally, supplementation of mice with milk-derived HA resulted in the release of the homologous murine beta-defensin through CD44- and TLR4-dependent mechanisms (Hill et al. 2012). Fragmented HA of varying sizes has been shown to have protective effects on the intestinal epithelium through the induction of endogenous cell signaling. In a mouse model of DSS colitis, Zheng et al. demonstrated intraperitoneal administration of 750 kDa HA ameliorated disease through a TLR4-, cyclooxygenase-2 (COX-2)-, and prostaglandin E2 (PGE2)-dependent proliferative repair of the intestinal epithelium (Zheng, Riehl, and Stenson 2009). These effects were largely conveyed via peritoneal macrophage activation of endogenous HAS-associated synthesis of HMW-HA, which upon fragmentation, induced epithelial repair and proliferation. Interestingly, several groups (e.g., (Riehl et al. 2015, O'Neill 2009)) have demonstrated an apparent feed-forward characteristic of exogenous HA, whereby endogenous HA synthesis is upregulated in response to HA supplementation. A similar effect is thought to occur within the joints, where clinically significant benefits of locally injected HA far outlast the biological life of the administered polymers (Bowman et al. 2018). This endogenous response to exogenously administered HA may indicate a potential role for HA-mediated repair of the intestine during or following NEC-like intestinal inflammation. Riehl et al. subjected mice to full body radiation (Riehl, Foster, and Stenson 2012). Mice receiving intraperitoneal HA 750 kDa pretreatment retained nearly double the small intestinal crypts and were significantly protected from enterocyte apoptosis compared to control. These effects were mediated through TLR4- and COX-2-dependent signaling, as well as mesenchymal stem cell migration from the villi lamina propria to that of the crypts. Asari et al. (Asari, Kanemitsu, and Kurihara 2010) demonstrated protection of intestinal epithelium in a T helper cell (Th)-1-type autoimmune mouse model via oral treatment of 900 kDa HA. HA induced a reduction in systemic inflammation through upregulation of the anti-inflammatory cytokine, IL-10. Epithelial TLR4-dependent upregulation of suppressor of cytokine signaling 3 (SOCS3) and downregulation of pleiotrophin resulted in a reduction in inflammatory T cell lymphoaccumulation. Finally, HA can directly affect the intestinal barrier through increases in tight junction expression (Kim and de la Motte 2020), preventing bacterial translocation from the lumen (Kim et al. 2017), critical processes in the pathogenesis of NEC (Remon et al. 2015, Heida et al. 2017). Specifically, intestinal expression of zonula occludens-1 (ZO-1), claudin-3, claudin-4, and occludin have been demonstrated to increase following treatment with HA 35 kDa (Kim et al. 2017, Gunasekaran et al. 2019), while effects on claudin-2 levels have been mixed (Kessler et al. 2018, Gunasekaran et al. 2019). Kim *et al.* (Kim et al. 2017) treated mice with HA 35 kDa, demonstrating protection from *Citrobacter rodentium* infection via an upregulation in the critical tight junction protein, ZO-1. This alteration to intestinal barrier function resulted in reduced intestinal permeability and inhibition of bacterial translocation across the intestinal epithelium. The ability of HA to directly impact NEC processes has been demonstrated in a murine model of the disease. Gunasekaran *et al.* (Gunasekaran et al. 2019) established the efficacy of HA 35 kDa in reducing both NEC incidence and severity in the intraperitoneal dithizone/oral *Klebsiella pneumoniae* two-hit model of the disease (Zhang et al. 2012). Mouse pups (age P14–16) were treated with HA 35 kDa (15 mg/kg or 30 mg/kg) once per day for three days prior to the initiation of NEC. Proinflammatory cytokine (tumor necrosis factor-alpha [TNF-α], growth-regulated oncogene-alpha [GRO-α], interleukin-12p70 [IL-12p70], and IL-6) release was significantly reduced with HA 35 kDa treatment (either dose) compared to untreated NEC. In addition to upregulation of tight junction proteins, these changes likely bolstered the intestinal barrier in response to the pathological bolus of bacteria, reducing bacteremia. Accompanying substantially greater survival, the ileum of pups treated with HA 35 kDa showed significantly diminished severity of the disease. Taken together, HA may reduce pathogen binding to intestinal epithelial cells, upregulate secretion of peptides associated with host defense, and reduce inflammatory signaling, resulting in bolstered intestinal barrier function and a significant decline in bacterial adhesion, invasion, and translocation.

CONCLUSION

Increasing evidence suggests the ability of HA to enhance the development of the fetal and postnatal gut, as well as protect against a severe gastrointestinal illness such as NEC. Through a plethora of mechanisms, HA enhances the barrier function of the intestinal epithelium, inhibits growth and invasion of pathogenic luminal bacteria, and strengthens innate immune defenses against gastrointestinal infection among neonates. In addition, rodent models of gastrointestinal development have helped elucidate the critical functions of HA during embryogenesis and fetal gastrointestinal development. Human milk-derived HA, as well as supplemental fragmented HA, hold great promise in preventing enteric infections, especially in the preterm environment. Further studies are needed to further characterize the effects of HM-derived HA, as well as differentially fragmented HA, on the intestinal epithelium, pathogen-specific interactions, and the microbiome, specifically in the contexts of prematurity and NEC.

REFERENCES

Abo, H., Chassaing, B., Harusato, A., Quiros, M., Brazil, J. C., Ngo, V. L., Viennois, E., Merlin, D., Gewirtz, A. T., Nusrat, A. & Denning, T. L. (2020). "Erythroid differentiation regulator-1 induced by microbiota in early life drives intestinal stem cell proliferation and regeneration." *Nat Commun*, 11 (1), 513. doi: 10.1038/s41467-019-14258-z.

Abrams, S. A., Schanler, R. J., Lee, M. L. & Rechtman, D. J. (2014). "Greater mortality and morbidity in extremely preterm infants fed a diet containing cow milk protein products." *Breastfeed Med*, 9 (6), 281-5. doi: 10.1089/bfm.2014.0024.

Ade-Ajayi, N., Spitz, L., Kiely, E., Drake, D. & Klein, N. (1996). "Intestinal glycosaminoglycans in neonatal necrotizing enterocolitis." *Brit J Surg*, 83, 415-418.

Armstrong, S. E. & Bell, D. R. (2002). "Measurement of high-molecular-weight hyaluronan in solid tissue using agarose gel electrophoresis." *Anal Biochem*, 308 (2), 255-64. doi: 10.1016/ s0003-2697(02)00239-7.

Asari, A., Kanemitsu, T. & Kurihara, H. (2010). "Oral administration of high molecular weight hyaluronan (900 kDa) controls immune system via Toll-like receptor 4 in the intestinal epithelium." *J Biol Chem*, 285 (32), 24751-8. doi: 10.1074/jbc.M110.104950.

Balogh, L., Polyak, A., Mathe, D., Kiraly, R., Thuroczy, J., Terez, M., Janoki, G., Ting, Y., Bucci, L. R. & Schauss, A. G. (2008). "Absorption, uptake and tissue affinity of high-molecular-weight hyaluronan after oral administration in rats and dogs." *J Agric Food Chem*, 56 (22), 10582-93. doi: 10.1021/jf8017029.

Barbosa de Souza, A., Vinícius Chaud, M., Francine Alves, T., Ferreira de Souza, J. & Andrade Santana, M. H. (2020). "Hyaluronic Acid in the Intestinal Tract: Influence of Structure, Rheology, and Mucoadhesion on the Intestinal Uptake in Rats." *Biomolecules*, 10 (10). doi: 10.3390/biom10101422.

Baricelli, J., Rocafull, M. A., Vázquez, D., Bastidas, B., Báez-Ramirez, E. & Thomas, L. E. (2015). "β-defensin-2 in breast milk displays a broad antimicrobial activity against pathogenic bacteria." *J Pediatr (Rio J)*, 91 (1), 36-43. doi: 10.1016/j.jped.2014.05.006.

Barthe, L., Woodley, J. & Houin, G. (1999). "Gastrointestinal absorption of drugs: methods and studies." *Fundam Clin Pharmacol*, 13 (2), 154-68. doi: 10.1111/j.1472-8206.1999.tb00334.x.

Battersby, A. J. & Gibbons, D. L. (2013). "The gut mucosal immune system in the neonatal period." *Pediatr Allergy Immunol*, 24 (5), 414-21. doi: 10.1111/pai.12079.

Bistoletti, M., Bosi, A., Caon, I., Chiaravalli, A. M., Moretto, P., Genoni, A., Moro, E., Karousou, E., Viola, M., Crema, F., Baj, A., Passi, A., Vigetti, D. & Giaroni, C. (2020). "Involvement of hyaluronan in the adaptive changes of the rat small intestine neuromuscular function after ischemia/reperfusion injury." *Sci Rep*, 10 (1), 11521. doi: 10. 1038/s41598-020-67876-9.

Bode, Lars, Camilla Salvestrini, Pyong Woo Park, Jin-Ping Li, Jeffrey D. Esko, Yu Yamaguchi, Simon Murch. & Hudson H. Freeze. (2008). "Heparan sulfate and syndecan-1 are essential in maintaining murine and human intestinal epithelial barrier function."

The Journal of Clinical Investigation, 118 (1), 229-238. doi: 10.1172/JCI32335.

Bowman, S., Awad, M. E., Hamrick, M. W., Hunter, M. & Fulzele, S. (2018). "Recent advances in hyaluronic acid based therapy for osteoarthritis." Clin Transl Med, 7 (1), 6. doi: 10.1186/s40169-017-0180-3.

Burge, K., Gunasekaran, A., Eckert, J. & Chaaban, H. (2019). "Curcumin and Intestinal Inflammatory Diseases: Molecular Mechanisms of Protection." Int J Mol Sci, 20 (1912). doi: 10.3390/ijms20081912.

Camenisch, T. D., Spicer, A. P., Brehm-Gibson, T., Biesterfeldt, J., Augustine, M. L., Calabro, A., Jr. Kubalak, S., Klewer, S. E. & McDonald, J. A. (2000). "Disruption of hyaluronan synthase-2 abrogates normal cardiac morphogenesis and hyaluronan-mediated transformation of epithelium to mesenchyme." J Clin Invest, 106 (3), 349-60. doi: 10.1172/jci10272.

Carlson, G. A., Dragoo, J. L., Samimi, B., Bruckner, D. A., Bernard, G. W., Hedrick, M. & Benhaim, P. (2004). "Bacteriostatic properties of biomatrices against common orthopaedic pathogens." Biochem Biophys Res Commun, 321 (2), 472-8. doi: 10.1016/j.bbrc.2004.06.165.

Chin, A. M., Hill, D. R., Aurora, M. & Spence, J. R. (2017). "Morphogenesis and maturation of the embryonic and postnatal intestine." Semin Cell Dev Biol, 66, 81-93. doi: 10.1016/j.semcdb.2017.01.011.

Clevers, H. C. & Bevins, C. L. (2013). "Paneth cells: maestros of the small intestinal crypts." Annu Rev Physiol, 75, 289-311. doi: 10.1146/annurev-physiol-030212-183744.

Coppa, G. V., Facinelli, B., Magi, G., Marini, E., Zampini, L., Mantovani, V., Galeazzi, T., Padella, L., Marchesiello, R. L., Santoro, L., Coscia, A., Peila, C., Volpi, N. & Gabrielli, O. (2016). "Human milk glycosaminoglycans inhibit in vitro the adhesion of Escherichia coli and Salmonella fyris to human intestinal cells." Pediatr Res, 79 (4), 603-7. doi: 10.1038/pr.2015.262.

Coppa, G. V., Gabrielli, O., Bertino, E., Zampini, L., Galeazzi, T., Padella, L., Santoro, L., Marchesiello, R. L., Galeotti, F., Maccari, F. & Volpi, N. (2013). "Human milk glycosaminoglycans: the state of the art and future perspectives." *Ital J Pediatr*, 39 (2). doi: 10.1186/1824-7288-39-2.

Coppa, G. V., Gabrielli, O., Buzzega, D., Zampini, L., Galeazzi, T., Maccari, F., Bertino, E. & Volpi, N. (2011). "Composition and structure elucidation of human milk glycosaminoglycans." *Glycobiology*, 21 (3), 295-303. doi: 10.1093/glycob/cwq164.

Corr, S. C., Gahan, C. C. & Hill, C. (2008). "M-cells: origin, morphology and role in mucosal immunity and microbial pathogenesis." *FEMS Immunol Med Microbiol*, 52 (1), 2-12. doi: 10.1111/j.1574-695X.2007.00359.x.

Cummins, A. G., Catto-Smith, A. G., Cameron, D. J., Couper, R. T., Davidson, G. P., Day, A. S., Hammond, P. D., Moore, D. J. & Thompson, F. M. (2008). "Crypt fission peaks early during infancy and crypt hyperplasia broadly peaks during infancy and childhood in the small intestine of humans." *J Pediatr Gastroenterol Nutr*, 47 (2), 153-7. doi: 10.1097/MPG.0b013e3181604d27.

Cyphert, J. M., Trempus, C. S. & Garantziotis, S. (2015). "Size Matters: Molecular Weight Specificity of Hyaluronan Effects in Cell Biology." *Int J Cell Biol*, 563818. doi: 10.1155/2015/563818.

Dahl, L. B., Dahl, I. M. & Børresen, A. L. (1986). "The molecular weight of sodium hyaluronate in amniotic fluid." *Biochem Med Metab Biol*, 35 (2), 219-26. doi: 10.1016/0885-4505(86)90077-0.

De Plaen, I. G. (2013). "Inflammatory signaling in necrotizing enterocolitis." *Clin Perinatol*, 40 (1), 109-24. doi: 10.1016/j.clp.2012.12.008.

De Plaen, I. G., Liu, S. X., Tian, R., Neequaye, I., May, M. J., Han, X. B., Hsueh, W., Jilling, T., Lu, J. & Caplan, M. S. (2007). "Inhibition of nuclear factor-kappaB ameliorates bowel injury and prolongs survival in a neonatal rat model of necrotizing enterocolitis." *Pediatr Res*, 61 (6), 716-21. doi: 10.1203/pdr.0b013e3180534219.

de Santa Barbara, P., van den Brink, G. R. & Roberts, D. J. (2003). "Development and differentiation of the intestinal epithelium." *Cell Mol Life Sci*, 60 (7), 1322-32. doi: 10.1007/s00018-003-2289-3.

de Souza, A. B., Chaud, M. V. & Santana, M. H. A. (2019). "Hyaluronic acid behavior in oral administration and perspectives for nanotechnology-based formulations: A review." *Carbohydr Polym*, 222, 115001. doi: 10.1016/j.carbpol.2019.115001.

Dehmer, J. J., Garrison, A. P., Speck, K. E., Dekaney, C. M., Van Landeghem, L., Sun, X., Henning, S. J. & Helmrath, M. A. (2011). "Expansion of intestinal epithelial stem cells during murine development." *PLoS One*, 6 (11), e27070. doi: 10.1371/journal.pone.0027070.

Denning, N. L. & Prince, J. M. (2018). "Neonatal intestinal dysbiosis in necrotizing enterocolitis." *Mol Med*, 24 (1), 4. doi: 10.1186/s10020-018-0002-0.

Dimmitt, R. A., Staley, E. M., Chuang, G., Tanner, S. M., Soltau, T. D. & Lorenz, R. G. (2010). "Role of postnatal acquisition of the intestinal microbiome in the early development of immune function." *J Pediatr Gastroenterol Nutr*, 51 (3), 262-73. doi: 10.1097/MPG.0b013e3181e1a114.

Ebid, R., Lichtnekert, J. & Anders, H. J. (2014). "Hyaluronan is not a ligand but a regulator of toll-like receptor signaling in mesangial cells: role of extracellular matrix in innate immunity." *ISRN Nephrol*, 714081. doi: 10.1155/2014/714081.

Fallacara, A., Baldini, E., Manfredini, S. & Vertuani, S. (2018). "Hyaluronic Acid in the Third Millennium." *Polymers (Basel)*, 10 (7). doi: 10.3390/polym10070701.

Filpa, V., Bistoletti, M., Caon, I., Moro, E., Grimaldi, A., Moretto, P., Baj, A., Giron, M. C., Karousou, E., Viola, M., Crema, F., Frigo, G., Passi, A., Giaroni, C. & Vigetti, D. (2017). "Changes in hyaluronan deposition in the rat myenteric plexus after experimentally-induced colitis." *Sci Rep*, 7 (1), 17644. doi: 10.1038/s41598-017-18020-7.

Good, M., Sodhi, C. P., Yamaguchi, Y., Jia, H., Lu, P., Fulton, W. B., Martin, L. Y., Prindle, T., Nino, D. F., Zhou, Q., Ma, C., Ozolek, J. A., Buck, R. H., Goehring, K. C. & Hackam, D. J. (2016). "The human milk oligosaccharide 2'-fucosyllactose attenuates the severity of experimental necrotising enterocolitis by enhancing mesenteric perfusion in the neonatal intestine." *Br J Nutr*, 116 (7), 1175-1187. doi: 10.1017/s0007114516002944.

Grosse, A. S., Pressprich, M. F., Curley, L. B., Hamilton, K. L., Margolis, B., Hildebrand, J. D. & Gumucio, D. L. (2011). "Cell dynamics in fetal intestinal epithelium: implications for intestinal growth and morphogenesis." *Development*, 138 (20), 4423-32. doi: 10.1242/dev.065789.

Grulee, C. G., Sanford, H. N. & Herron, P. H. (1934). "Breast and artificial feeding: Influence on morbidity and mortality of twenty thousand infants." *JAMA: The Journal of the American Medical Association*, 103 (10), 735-738.

Gunasekaran, A., Eckert, J., Burge, K., Zheng, W., Yu, Z., Kessler, S., de la Motte, C. & Chaaban, H. (2019). "Hyaluronan 35 kDa enhances epithelial barrier function and protects against the development of murine necrotizing enterocolitis." *Pediatr Res*. doi: 10.1038/s41390-019-0563-9.

Hao, M. M., Foong, J. P., Bornstein, J. C., Li, Z. L., Vanden Berghe, P. & Boesmans, W. (2016). "Enteric nervous system assembly: Functional integration within the developing gut." *Dev Biol*, 417 (2), 168-81. doi: 10.1016/j.ydbio.2016.05.030.

Hatch, J. & Mukouyama, Y. S. (2015). "Spatiotemporal mapping of vascularization and innervation in the fetal murine intestine." *Dev Dyn*, 244 (1), 56-68. doi: 10.1002/dvdy.24178.

Heida, F. H., Harmsen, H. J., Timmer, A., Kooi, E. M., Bos, A. F. & Hulscher, J. B. (2017). "Identification of bacterial invasion in necrotizing enterocolitis specimens using fluorescent *in situ* hybridization." *J Perinatol*, 37 (1), 67-72. doi: 10.1038/jp.2016.165.

Hill, D. R., Kessler, S. P., Rho, H. K., Cowman, M. K. & de la Motte, C. A. (2012). "Specific-sized hyaluronan fragments promote expression of human beta-defensin 2 in intestinal epithelium." *J Biol Chem*, 287 (36), 30610-24. doi: 10.1074/jbc.M112.356238.

Hill, D. R., Rho, H. K., Kessler, S. P., Amin, R., Homer, C. R., McDonald, C., Cowman, M. K. & de la Motte, C. A. (2013). "Human milk hyaluronan enhances innate defense of the intestinal epithelium." *J Biol Chem*, 288 (40), 29090-104. doi: 10.1074/jbc.M113.468629.

Howie, P. W., Forsyth, J. S., Ogston, S. A., Clark, A. & Florey, C. D. (1990). "Protective effect of breast feeding against infection." *Bmj*, 300 (6716), 11-6. doi: 10.1136/bmj.300.6716.11.

Huang, P. M., Syrkina, O., Yu, L., Dedaj, R., Zhao, H., Shiedlin, A., Liu, Y. Y., Garg, H., Quinn, D. A. & Hales, C. A. (2010). "High MW hyaluronan inhibits smoke inhalation-induced lung injury and improves survival." *Respirology*, 15 (7), 1131-9. doi: 10.1111/j.1440-1843.2010.01829.x.

Jackson, D. G. (2009). "Immunological functions of hyaluronan and its receptors in the lymphatics." *Immunol Rev*, 230 (1), 216-31. doi: 10.1111/j.1600-065X.2009.00803.x.

Jiang, D., Liang, J. & Noble, P. W. (20110. "Hyaluronan as an immune regulator in human diseases." *Physiol Rev*, 91 (1), 221-64. doi: 10.1152/physrev.00052.2009.

Karamanos, N. K., Piperigkou, Z., Theocharis, A. D., Watanabe, H., Franchi, M., Baud, S., Brézillon, S., Götte, M., Passi, A., Vigetti, D., Ricard-Blum, S., Sanderson, R. D., Neill, T. & Iozzo, R. V. (2018). "Proteoglycan Chemical Diversity Drives Multifunctional Cell Regulation and Therapeutics." *Chem Rev*, 118 (18), 9152-9232. doi: 10.1021/acs.chemrev.8b00354.

Kessler, S. P., Obery, D. R., Nickerson, K. P., Petrey, A. C., McDonald, C. & de la Motte, C. A. (2018). "Multifunctional Role of 35 Kilodalton Hyaluronan in Promoting Defense of the Intestinal Epithelium." *J Histochem Cytochem*, 66 (4), 273-287. doi: 10.1369/0022155417746775.

Kim, Y. & de la Motte, C. A. (2020). "The Role of Hyaluronan Treatment in Intestinal Innate Host Defense." *Front Immunol*, 11, 569. doi: 10.3389/fimmu.2020.00569.

Kim, Y., Kessler, S. P., Obery, D. R., Homer, C. R., McDonald, C. & de la Motte, C. A. (2017). "Hyaluronan 35 kDa treatment protects mice from Citrobacter rodentium infection and induces epithelial tight junction protein ZO-1 *in vivo*." *Matrix Biol*, 62, 28-39. doi: 10.1016/j.matbio.2016.11.001.

Kim, Y., West, G. A., Ray, G., Kessler, S. P., Petrey, A. C., Fiocchi, C., McDonald, C., Longworth, M. S., Nagy, L. E. & de la Motte, C. A. (2018). "Layilin is critical for mediating hyaluronan 35kDa-induced intestinal epithelial tight junction protein ZO-1 *in vitro* and *in vivo*." *Matrix Biol*, 66, 93-109. doi: 10.1016/j.matbio.2017.09.003.

Kimura, M., Maeshima, T., Kubota, T., Kurihara, H., Masuda, Y. & Nomura, Y. (2016). "Absorption of Orally Administered Hyaluronan." *J Med Food*, 19 (12), 1172-1179. doi: 10.1089/jmf.2016.3725.

Kogan, G., Soltés, L., Stern, R. & Mendichi, R. (2007). "Hyaluronic acid: A biopolymer with versatile physico-chemical and biological properties." In *Handbook of Polymer Research: Monomers,*

Oligomers, Polymers and Composites, edited by R.A. Pethrich, B. Antonio and G.E. Zaikov, 393-439. New York, NY: Nova Science Publishers Inc.

Koropatkin, Nicole M., Elizabeth A. Cameron. & Eric C. Martens. (2012). "How glycan metabolism shapes the human gut microbiota." *Nature Reviews Microbiology*, 10 (5), 323-335. doi: 10.1038/nrmicro2746.

Krappmann, D., Wegener, E., Sunami, Y., Esen, M., Thiel, A., Mordmuller, B. & Scheidereit, C. (2004). "The IkappaB kinase complex and NF-kappaB act as master regulators of lipopolysaccharide-induced gene expression and control subordinate activation of AP-1." *Mol Cell Biol*, 24 (14), 6488-500. doi: 10.1128/mcb.24.14.6488-6500.2004.

Laurent, T. (1989). "The biology of hyaluronan. Introduction." *Ciba Found Symp*, 143, 1-20.

Lee, Yonghyun, Kohei Sugihara, Merritt G. Gillilland, Sangyong Jon, Nobuhiko Kamada. & James J. Moon. (2020). "Hyaluronic acid–bilirubin nanomedicine for targeted modulation of dysregulated intestinal barrier, microbiome and immune responses in colitis." *Nature Materials* 19 (1):118-126. doi: 10.1038/s41563-019-0462-9.

Li, X. G., Wang, Z., Chen, R. Q., Fu, H. L., Gao, C. Q., Yan, H. C., Xing, G. X. & Wang, X. Q. (2018). "LGR5 and BMI1 Increase Pig Intestinal Epithelial Cell Proliferation by Stimulating WNT/β-Catenin Signaling." *Int J Mol Sci*, 19 (4). doi: 10.3390/ijms19041036.

Lu, P., Sodhi, C. P. & Hackam, D. J. (2014). "Toll-like receptor regulation of intestinal development and inflammation in the pathogenesis of necrotizing enterocolitis." *Pathophysiology*, 21 (1), 81-93. doi: 10.1016/j.pathophys.2013.11.007.

Maccari, F., Mantovani, V., Gabrielli, O., Carlucci, A., Zampini, L., Galeazzi, T., Galeotti, F., Coppa, G. V. & Volpi, N. (2016). "Metabolic fate of milk glycosaminoglycans in breastfed and formula fed newborns." *Glycoconj J*, 33 (2), 181-8. doi: 10.1007/s10719-016-9655-5.

Mallow, E. B., Harris, A., Salzman, N., Russell, J. P., DeBerardinis, R. J., Ruchelli, E. & Bevins, C. L. (1996). "Human enteric defensins. Gene structure and developmental expression." *J Biol Chem*, 271 (8), 4038-45. doi: 10.1074/jbc.271.8.4038.

Managlia, E., Liu, S. X. L., Yan, X., Tan, X. D., Chou, P. M., Barrett, T. A. & De Plaen, I. G. (2019). "Blocking NF-kappaB Activation in Ly6c(+) Monocytes Attenuates Necrotizing Enterocolitis." *Am J Pathol*, 189 (3), 604-618. doi: 10.1016/j.ajpath.2018.11.015.

Mara, M. A., Good, M. & Weitkamp, J. H. (2018). "Innate and adaptive immunity in necrotizing enterocolitis." *Semin Fetal Neonatal Med*, 23 (6), 394-399. doi: 10.1016/j.siny.2018.08.002.

Markel, T. A., Crisostomo, P. R., Wairiuko, G. M., Pitcher, J., Tsai, B. M. & Meldrum, D. R. (2006). "Cytokines in necrotizing enterocolitis." *Shock*, 25 (4), 329-37. doi: 10.1097/01.shk.0000192126.33823.87.

McHugh, K. M. (1996). "Molecular analysis of gastrointestinal smooth muscle development." *J Pediatr Gastroenterol Nutr*, 23 (4), 379-94. doi: 10.1097/00005176-199611000-00001.

Monslow, J., Govindaraju, P. & Puré, E. (2015). "Hyaluronan - a functional and structural sweet spot in the tissue microenvironment." *Front Immunol*, 6, 231. doi: 10.3389/fimmu.2015.00231.

Moore, S. A., Nighot, P., Reyes, C., Rawat, M., McKee, J., Lemon, D., Hanson, J. & Ma, T. Y. (2016). "Intestinal barrier dysfunction in human necrotizing enterocolitis." *J Pediatr Surg*, 51 (12), 1907-1913. doi: 10.1016/j.jpedsurg.2016.09.011.

Neal, M. D., Leaphart, C., Levy, R., Prince, J., Billiar, T. R., Watkins, S., Li, J., Cetin, S., Ford, H., Schreiber, A. & Hackam, D. J. (2006). "Enterocyte TLR4 mediates phagocytosis and translocation of bacteria across the intestinal barrier." *J Immunol*, 176 (5), 3070-9. doi: 10.4049/jimmunol.176.5.3070.

Neu, J. & Walker, W. A. (2011). "Necrotizing enterocolitis." *N Engl J Med*, 364 (3), 255-64. doi: 10.1056/NEJMra1005408.

Neutra, M. R. (1998). "Current concepts in mucosal immunity. V Role of M cells in transepithelial transport of antigens and pathogens to the mucosal immune system." *Am J Physiol*, 274 (5), G785-91. doi: 10.1152/ajpgi.1998.274.5.G785.

O'Neill, L. A. (2009. "A feed-forward loop involving hyaluronic acid and toll-like receptor-4 as a treatment for colitis?" *Gastroenterology*, 137 (6), 1889-91. doi: 10.1053/j.gastro.2009.10.015.

Oe, M., Mitsugi, K., Odanaka, W., Yoshida, H., Matsuoka, R., Seino, S., Kanemitsu, T. & Masuda, Y. (2014). "Dietary hyaluronic acid migrates into the skin of rats." *Scientific World Journal*, 2014, 378024. doi: 10.1155/2014/378024.

Oe, M., Tashiro, T., Yoshida, H., Nishiyama, H., Masuda, Y., Maruyama, K., Koikeda, T., Maruya, R. & Fukui, N. (2016). "Oral hyaluronan relieves knee pain: a review." *Nutr J*, 15, 11. doi: 10.1186/s12937-016-0128-2.

Pammi, M., Cope, J., Tarr, P. I., Warner, B. B., Morrow, A. L., Mai, V., Gregory, K. E., Kroll, J. S., McMurtry, V., Ferris, M. J., Engstrand, L., Lilja, H. E., Hollister, E. B., Versalovic, J. & Neu, J. (2017). "Intestinal dysbiosis in preterm infants preceding necrotizing enterocolitis: a systematic review and meta-analysis." *Microbiome*, 5 (1), 31. doi: 10.1186/s40168-017-0248-8.

Papakonstantinou, E., Roth, M. & Karakiulakis, G. (2012). "Hyaluronic acid: A key molecule in skin aging." *Dermatoendocrinol*, 4 (3), 253-8. doi: 10.4161/derm.21923.

Pirnazar, P., Wolinsky, L., Nachnani, S., Haake, S., Pilloni, A. & Bernard, G. W. (1999). "Bacteriostatic effects of hyaluronic acid." *J Periodontol*, 70 (4), 370-4. doi: 10.1902/jop.1999.70.4.370.

Prehm, P. (1990). "Release of hyaluronate from eukaryotic cells." *Biochem J*, 267 (1), 185-9. doi: 10.1042/bj2670185.

Remon, J. I., Amin, S. C., Mehendale, S. R., Rao, R., Luciano, A. A., Garzon, S. A. & Maheshwari, A. (2015). "Depth of bacterial invasion in resected intestinal tissue predicts mortality in surgical necrotizing enterocolitis." *J Perinatol*, 35 (9), 755-62. doi: 10.1038/jp.2015.51.

Riehl, T. E., Ee, X. & Stenson, W. F. (2012). "Hyaluronic acid regulates normal intestinal and colonic growth in mice." *Am J Physiol Gastrointest Liver Physiol*, 303 (3), G377-88. doi: 10.1152/ajpgi.00034.2012.

Riehl, T. E., Foster, L. & Stenson, W. F. (2012). "Hyaluronic acid is radioprotective in the intestine through a TLR4 and COX-2-mediated mechanism." *Am J Physiol Gastrointest Liver Physiol*, 302 (3), G309-16. doi: 10.1152/ajpgi.00248.2011.

Riehl, T. E., Santhanam, S., Foster, L., Ciorba, M. & Stenson, W. F. (2015). "CD44 and TLR4 mediate hyaluronic acid regulation of Lgr5+ stem cell proliferation, crypt fission, and intestinal growth in postnatal and adult mice." *Am J Physiol Gastrointest Liver Physiol*, 309 (11), G874-87. doi: 10.1152/ajpgi.00123.2015.

Round, J. L. & Mazmanian, S. K. (2009). "The gut microbiota shapes intestinal immune responses during health and disease." *Nat Rev Immunol*, 9 (5), 313-23. doi: 10.1038/nri2515.

Rubas, Werner. & George M. Grass. (1991). "Gastrointestinal lymphatic absorption of peptides and proteins." *Advanced Drug Delivery Reviews*, 7 (1), 15-69. doi: 10.1016/0169-409x(91)90047-g.

Ruppert, S. M., Hawn, T. R., Arrigoni, A., Wight, T. N. & Bollyky, P. L. (2014). "Tissue integrity signals communicated by high-molecular weight hyaluronan and the resolution of inflammation." *Immunol Res*, 58 (2-3), 186-92. doi: 10.1007/s12026-014-8495-2.

Samuels, N., van de Graaf, R. A., de Jonge, R. C. J., Reiss, I. K. M. & Vermeulen, M. J. (2017). "Risk factors for necrotizing enterocolitis in neonates: a systematic review of prognostic studies." *BMC Pediatr*, 17 (1), 105. doi: 10.1186/s12887-017-0847-3.

Scheibner, K. A., Lutz, M. A., Boodoo, S., Fenton, M. J., Powell, J. D. & Horton, M. R. (2006). "Hyaluronan fragments act as an endogenous danger signal by engaging TLR2." *J Immunol*, 177 (2), 1272-81. doi: 10.4049/jimmunol.177.2.1272.

Shyer, A. E., Tallinen, T., Nerurkar, N. L., Wei, Z., Gil, E. S., Kaplan, D. L., Tabin, C. J. & Mahadevan, L. (2013). "Villification: how the gut gets its villi." *Science*, 342 (6155), 212-8. doi: 10.1126/science.1238842.

Sisk, P. M., Lovelady, C. A., Dillard, R. G., Gruber, K. J. & O'Shea, T. M. (2007). "Early human milk feeding is associated with a lower risk of necrotizing enterocolitis in very low birth weight infants." *J Perinatol*, 27 (7), 428-33. doi: 10.1038/sj.jp.7211758.

Sivakumar, A., Mahadevan, A., Lauer, M. E., Narvaez, R. J., Ramesh, S., Demler, C. M., Souchet, N. R., Hascall, V. C., Midura, R. J., Garantziotis, S., Frank, D. B., Kimata, K. & Kurpios, N. A. (2018). "Midgut Laterality Is Driven by Hyaluronan on the Right." *Dev Cell*, 46 (5), 533-551.e5. doi: 10.1016/j.devcel.2018.08.002.

Slevin, M., Krupinski, J., Gaffney, J., Matou, S., West, D., Delisser, H., Savani, R. C. & Kumar, S. (2007). "Hyaluronan-mediated angiogenesis in vascular disease: uncovering RHAMM and CD44 receptor signaling pathways." *Matrix Biol*, 26 (1), 58-68. doi: 10.1016/j.matbio.2006.08.261.

Slevin, M., Kumar, S. & Gaffney, J. (2002). "Angiogenic oligosaccharides of hyaluronan induce multiple signaling pathways affecting vascular endothelial cell mitogenic and wound healing responses." *J Biol Chem*, 277 (43), 41046-59. doi: 10.1074/jbc.M109443200.

Soltés, L., Mendichi, R., Kogan, G., Schiller, J., Stankovska, M. & Arnhold, J. (2006). "Degradative action of reactive oxygen species on hyaluronan." *Biomacromolecules*, 7 (3), 659-68. doi: 10.1021/bm050867v.

Sprangers, J., Zaalberg, I. C. & Maurice, M. M. (2021). "Organoid-based modeling of intestinal development, regeneration, and repair." *Cell Death Differ*, 28 (1), 95-107. doi: 10.1038/s41418-020-00665-z.

Stern, R., Asari, A. A. & Sugahara, K. N. (2006). "Hyaluronan fragments: an information-rich system." *Eur J Cell Biol*, 85 (8), 699-715. doi: 10.1016/j.ejcb.2006.05.009.

Tanner, S. M., Berryhill, T. F., Ellenburg, J. L., Jilling, T., Cleveland, D. S., Lorenz, R. G. & Martin, C. A. (2015). "Pathogenesis of necrotizing enterocolitis: modeling the innate immune response." *Am J Pathol*, 185 (1), 4-16. doi: 10.1016/j.ajpath.2014.08.028.

Thoma, C. (2019). "Preventing brain damage in necrotizing enterocolitis." *Nat Rev Gastroenterol Hepatol*, 16 (2), 75. doi: 10.1038/s41575-019-0107-0.

Turner, J. R. (2009). "Intestinal mucosal barrier function in health and disease." *Nat Rev Immunol*, 9 (11), 799-809. doi: 10.1038/nri2653.

Udall, J. N., Pang, K., Fritze, L., Kleinman, R. & Walker, W. A. (1981). "Development of gastrointestinal mucosal barrier. I. The effect of age on intestinal permeability to macromolecules." *Pediatr Res*, 15 (3), 241-4.

Volpi, N., Schiller, J., Stern, R. & Soltés, L. (2009). "Role, metabolism, chemical modifications and applications of hyaluronan." *Curr Med Chem*, 16 (14), 1718-45. doi: 10.2174/092986709788186138.

Wang, C., Lang, Y., Li, Q., Jin, X., Li, G. & Yu, G. (2018). "Glycosaminoglycanomic profiling of human milk in different stages of lactation by liquid chromatography-tandem mass spectrometry." *Food Chem*, 258, 231-236. doi: 10.1016/j.foodchem.2018.03.076.

Watkins, D. J. & Besner, G. E. (2013). "The role of the intestinal microcirculation in necrotizing enterocolitis." *Semin Pediatr Surg*, 22 (2), 83-7. doi: 10.1053/j.sempedsurg.2013.01.004.

Yazji, I., Sodhi, C. P., Lee, E. K., Good, M., Egan, C. E., Afrazi, A., Neal, M. D., Jia, H., Lin, J., Ma, C., Branca, M. F., Prindle, T., Richardson, W. M., Ozolek, J., Billiar, T. R., Binion, D. G., Gladwin, M. T. & Hackam, D. J. (2013). "Endothelial TLR4 activation impairs intestinal microcirculatory perfusion in necrotizing enterocolitis via eNOS-NO-nitrite signaling." *Proc Natl Acad Sci U S A*, 110 (23), 9451-6. doi: 10.1073/pnas.1219997110.

Yuan, H., Amin, R., Ye, X., de la Motte, C. A. & Cowman, M. K. (2015). "Determination of hyaluronan molecular mass distribution in human breast milk." *Anal Biochem*, 474, 78-88. doi: 10.1016/j.ab.2014.12.020.

Zhang, C., Sherman, M. P., Prince, L. S., Bader, D., Weitkamp, J. H., Slaughter, J. C. & McElroy, S. J. (2012). "Paneth cell ablation in the presence of Klebsiella pneumoniae induces necrotizing enterocolitis (NEC)-like injury in the small intestine of immature mice." *Dis Model Mech*, 5 (4), 522-32. doi: 10.1242/dmm.009001.

Zheng, L., Riehl, T. E. & Stenson, W. F. (2009). "Regulation of colonic epithelial repair in mice by Toll-like receptors and hyaluronic acid." *Gastroenterology*, 137 (6), 2041-51. doi: 10.1053/j.gastro.2009.08.055.

ABOUT THE EDITOR

Vittorio Unfer
A.G.UN.CO
Obstetrics and Gynecology Center,
Rome, Italy

Vittorio Unfer (Rome, October 12, 1963) is a gynecologist, researcher, entrepreneur, and university professor. He has long been engaged in various researches on the etiopathogenesis and treatment of Polycystic Ovary Syndrome (PCOS). He is also responsible for important contributions on various therapies for infertility of both sexes.

In this regard, Vittorio Unfer was the first promoter of the use of myo-inositol in various therapeutic areas, paving the way for new approaches that have met with great interest in the scientific world. He is the author of numerous scientific publications and various international patents.

Degree in Medicine and Surgery from the University of Rome "Sapienza" (1991).

Specialization in Obstetrics and Gynecology at the II Institute of Obstetrics and Gynecology Clinic of the Policlinico Umberto I of Rome (1995). Since 1995 he has followed more than 25 refresher courses in the profession.

He is currently a professor in the Department of Experimental Medicine of the "Sapienza" University of Rome. Since 1999 he has held various positions as a university professor: "Legal Medicine" at the Degree Course in Obstetrician at the University of Rome "Sapienza"; "Health Statistics" at the Degree Course in Obstetrician at the University of Rome "Sapienza"; "Nutrition, nutrition and food pathology in pregnancy" at the School of Specialization in Gynecology and Obstetrics of the University of Perugia; "Endocrinology of childhood, adolescence and women" at the II level University Master at the School of Specialization in Endocrinology, at the University of Messina; "Unconventional medicines in obstetric and nursing care and obstetric autonomy in the management of pregnancy, childbirth and low-risk childbirth" at the Faculty of Medicine and Surgery of the University of Modena and Reggio Emilia.

He has participated as speaker at numerous national and international conferences in the field of endocrinology, gynecology and obstetrics.

Since 2008 he has been Vice President of SIFIP, the Italian Society of Phytotherapy and Supplements in Psychopathology. Since 2005 he has been a member of the Verduci ECM Scientific Committee. Since 2010 he has been a founding member of ISCHOM, the international chocolate and cocoa society in medicine.

Since 2016 he has been Editor in Chief of the scientific journal "International Journal of Medical Device and Adjuvant Treatments" (IJMDAT).

Since 2017 he has been part of the Board of Directors of the International Society of Dietary Supplements and Phytotherapy (ISDSP), whose mission is to aggregate specialists of international level to promote the correct use of food supplements and phytotherapy in the daily diet and in clinical practice. ISDSP is the result of the evolution of the Italian Society of Phytotherapy and Supplements in

Obstetrics and Gynecology (SIFIOG) of which Vittorio Unfer was President.

Since 2018 he is Editor in Chief of the Italian scientific journal Phitogynecea.

Inventor and owner of 16 national and international patents.

INDEX

#

750 kDa HA ameliorated disease, 155

A

absorbed, 11, 12, 144, 146, 147, 148
acid, 12, 144, 150
ADAMTS, 110, 112, 120, 126, 130, 134, 137
aggrecan, 109, 112, 122, 129, 132, 133, 135, 139, 140
angiogenesis, 13, 24, 42, 43, 60, 87, 94, 103, 109, 115, 119, 120, 121, 127, 133, 136, 138, 145, 150, 167
ART, 73, 83, 84, 85, 88, 91, 97, 101
autoimmune, 155

B

bacteriostatic and bactericidal effects, 154
bacteriostatic and bactericidal properties of HA, 153
blastocyst implantation, 23, 53, 58, 100
blastocyst(s), xi, 23, 25, 26, 27, 28, 37, 40, 43, 53, 58, 73, 78, 86, 87, 88, 89, 90, 93, 94, 95, 97, 99, 100

C

cardiac cushion(s), 108, 111, 112, 113
cardiac morphogenesis, 106, 111, 112, 113, 159
catabolism, 6, 140
CD44, 7, 8, 13, 14, 15, 17, 18, 19, 21, 24, 25, 26, 27, 29, 30, 36, 39, 41, 42, 44, 46, 49, 50, 56, 60, 73, 76, 86, 87, 108, 110, 113, 115, 121, 124, 126, 127, 128, 133, 135, 136, 138, 139, 145, 151, 155, 166, 167
cervix, xi, 19, 31, 32, 33, 36, 40, 41, 53, 56, 57, 59, 63, 69, 70, 74, 78, 84
cleft palate, 122
cor triatriatum, 112, 127, 136
craniofacial, 121, 122, 127

D

decidualization, 27, 28, 46, 62
depolymerization of HA, 144

developing embryo, 84, 105, 106, 118, 124, 150
development, x, xi, 6, 21, 25, 26, 27, 37, 38, 39, 41, 42, 43, 55, 56, 60, 84, 87, 91, 98, 99, 102, 105, 106, 107, 108, 109, 111, 112, 113, 115, 116, 117, 118, 119, 120, 121, 122, 123, 125, 126, 127, 128, 129, 130, 132, 133, 134, 135, 136, 137, 138, 139, 143, 147, 149, 151, 152, 154, 157, 160, 161, 162, 164, 165, 168
developmental, 77, 84, 87, 102, 105, 108, 111, 113, 119, 124, 127, 128, 129, 132, 135, 136, 137, 138, 139, 140, 141, 143, 146, 151, 153, 164
dorsal mesentery, 114, 116, 128
DSS colitis, 155

E

embryo development, 14, 23, 25, 26, 27, 44, 49, 60, 76, 84, 98, 102
embryo(s), 14, 15, 23, 25, 26, 27, 28, 29, 38, 41, 42, 44, 45, 49, 56, 60, 72, 73, 76, 78, 79, 82, 83, 84, 85, 86, 88, 89, 90, 91, 92, 93, 94, 95, 96, 97, 98, 99, 100, 101, 102, 103, 105, 106, 107, 108, 109, 110, 111, 112, 114, 116, 118, 119, 120, 121, 122, 124, 126, 135, 150
embryogenesis, 150
epithelial proliferation and apoptosis is dependent upon endogenous HA receptor binding of both CD44 and TLR4, 152
extracellular matrix (ECM), xi, 1, 3, 4, 7, 10, 15, 24, 27, 28, 29, 30, 31, 33, 34, 36, 37, 39, 42, 47, 53, 54, 56, 57, 58, 94, 105, 106, 107, 109, 110, 111, 112, 116, 121, 122, 124, 129, 144, 145, 150, 161

F

fetal development, ix, 41, 98, 105, 151
fetal-maternal interface, 34, 37

G

glycosaminoglycan(s) (GAG), 2, 4, 16, 31, 39, 40, 41, 42, 48, 53, 54, 55, 56, 58, 59, 60, 62, 63, 85, 98, 99, 103, 106, 107, 109, 116, 144, 147, 153, 157, 159, 160, 164
gut looping, 114, 132

H

HA, 2, 23, 144
HA 35 kDa in reducing both NEC incidence and severity, 156
HA concentration, 11, 57, 66, 73, 74, 148
HA in human milk (HM), 147, 153, 154, 157
HA polymer is synthesized, 144
HA polymer MW ranges, 145
HA receptors, 9, 24, 27, 95, 109, 145
HARE, 8, 16, 109
HAS, 5, 6, 25, 26, 31, 39, 86, 144, 155
HAS1, 5, 27, 107, 110, 120, 121, 134, 150, 151
HAS2, 5, 6, 24, 25, 26, 27, 29, 31, 58, 86, 107, 110, 111, 114, 118, 121, 122, 123, 134, 150
HAS3, 5, 24, 26, 39, 86, 107, 110, 134, 150
heart looping, 113
HETM, 91, 92, 94
high, xi, 2, 3, 4, 9, 10, 11, 12, 13, 17, 22, 24, 29, 36, 37, 41, 45, 47, 55, 56, 58, 65, 66, 67, 78, 83, 93, 95, 100, 102, 107, 108, 109, 118, 124, 144, 151, 157, 158, 162, 167

high molecular weight, 2, 4, 9, 10, 11, 12, 36, 56, 58, 78, 107, 118, 144, 158
high molecular weight hyaluronic acid (HMW-HA), xi, 2, 4, 7, 8, 9, 10, 11, 12, 23, 24, 25, 26, 27, 29, 30, 31, 32, 33, 34, 36, 37, 58, 59, 65, 72, 73, 75, 78, 85, 92, 95, 107, 108, 111, 116, 118, 143, 144, 145, 146, 147, 150, 151, 155
human β-defensin 2, 154
HYAL, 6, 107, 144
HYAL2, 6, 26, 107, 110, 112, 118, 122, 136
hyaluronan, ix, 1, 2, 3, 4, 5, 6, 7, 8, 9, 10, 13, 14, 15, 16, 17, 18, 19, 20, 21, 22, 24, 25, 29, 31, 37, 39, 40, 41, 42, 43, 44, 45, 47, 48, 49, 50, 54, 56, 57, 59, 60, 61, 62, 63, 71, 73, 78, 81, 82, 83, 85, 86, 87, 92, 93, 94, 96, 97, 98, 99, 100, 101, 102, 103,105, 107, 110, 114, 118, 125, 126, 127, 128, 129, 130, 131, 132, 133, 134, 135, 136, 137, 138, 139, 140, 141, 144, 157, 158, 159, 160, 161, 162, 163, 164, 165, 166, 167, 168, 169
hyaluronan synthase(s), 2, 5, 6, 16, 20, 21, 24, 31, 39, 42, 43, 48, 49, 63, 107, 110, 126, 134, 159
hyaluronic acid (HA), ix, x, xi, 1, 2, 3, 4, 5, 6, 7, 8, 9, 10, 11, 12, 13, 14, 15, 16, 18, 19, 20, 21, 23, 24, 25, 26, 27, 28, 29, 30, 31, 32, 33, 34, 36, 37, 40, 41, 46, 47, 48, 49, 50, 53, 54, 55, 56, 57, 58, 59, 60, 61, 62, 63, 65, 66, 67, 68, 69, 70, 71, 72, 73, 74, 75, 76, 77, 78, 79, 80, 81, 82, 83, 85, 86, 87, 88, 91, 92, 93, 94, 95, 99, 101, 102, 105, 106, 107, 108, 109, 110, 111, 112, 113, 114, 115, 117, 118, 119, 120, 121, 122, 123, 124, 125, 128, 132, 133, 137, 141, 143, 144, 145, 146, 147, 148, 150, 151, 152, 153, 157, 158, 159, 160, 161, 163, 164, 165, 166, 169
hyaluronic acid (HA) is a natural glycosaminoglycan, 144
hyaluronic acid in human milk, 147
hyaluronic acid in intestinal development, 148
hyaluronic acid is essential throughout embryogenesis and fetal developmen, 150
hyaluronidases, 2, 6, 7, 18, 32, 105, 107, 110, 112, 118, 124, 144

I

IL-10, 30, 39, 155
immune tolerance, xi, 23, 30, 37
immunity, 3, 13
immunity response, 3, 13
implantation, xi, 14, 23, 25, 26, 27, 28, 29, 37, 38, 39, 42, 43, 49, 50, 56, 72, 73, 76, 82, 83, 84, 85, 88, 89, 90, 91, 92, 93, 95, 96, 97, 98, 99, 100, 101, 102, 103, 105, 118, 124, 129
incidence of NEC, 147
intestinal barrier, 11, 18, 146, 153, 164, 165
intestinal morphogenesis, 106, 114
IVF, 43, 72, 73, 78, 83, 84, 89, 91, 93, 97, 98, 102

L

left-right asymmetry through a tumor necrosis factor-α-stimulated gene 6 (Tsg6)-dependent mechanism, 150
left-right organ asymmetry, 114
live birth(s), 71, 73, 83, 84, 90, 93, 95, 113
LMW-HA, xi, 4, 7, 8, 10, 11, 12, 24, 26, 30, 32, 33, 34, 36, 40, 143, 146, 147, 148

low molecular weight, xi, 2, 4, 9, 47, 107
low molecular weight hyaluronan, 2
LYVE-1, 8, 18, 20, 22, 27, 130, 131

M

M cells, 11, 144, 146, 147, 165
maternal-fetal interface, 28, 29
matrices rich in HMW-HA, 150
microbiome, 33, 47, 150, 153, 154, 157, 161, 164, 166
molecular weight(s), xi, 1, 3, 4, 5, 8, 9, 10, 11, 12, 15, 19, 29, 30, 36, 45, 74, 128, 132, 146, 160, 167
MW distribution, 148
myenteric neuron homeostasis, 152

N

NEC pathogenesis, 153
necrotizing enterocolitis (NEC), x, 143, 147, 152, 153, 157, 160, 161, 162, 164, 165, 166, 167, 168, 169
neonatal, 58, 61, 74, 79, 106, 123, 124, 131, 134, 143, 147, 150, 153, 157, 158, 160, 161, 165
neural crest cells, 121, 127, 129, 136, 149

O

odontogenesis, 121, 123
oocyte(s), 15, 24, 36, 38, 42, 43, 45, 48, 49, 50, 56, 62, 84, 88, 89, 90, 91, 92, 95, 96, 101, 108
oral HA, 146
oral HA fragment supplementation, 146
ovulation, 23, 24, 25, 28, 39, 49, 84, 86, 88, 102

P

P4, 34, 36
pathogenesis, 14, 87, 103, 153, 160, 164, 168
PEP-1 decreased, 151
PGRMC1, xi, 23, 34, 35, 36, 37, 44, 46, 47, 50
physiological pregnancy, ix, 23, 53, 54
Pitx2, 114, 116, 130, 134, 137, 140
polysaccharide(s), 1, 3, 4, 15, 18, 54
post-natal, 70, 105, 124
pregnancy, ix, xi, 23, 26, 28, 29, 30, 31, 32, 33, 34, 35, 36, 37, 39, 40, 41, 42, 43, 44, 45, 46, 47, 49, 50, 51, 53, 55, 56, 57, 58, 59, 60, 61, 62, 63, 65, 66, 67, 69, 70, 72, 73, 74, 75, 77, 79, 82, 83, 84, 85, 89, 92, 93, 95, 97, 98, 102, 103, 118, 130, 135
preventing bacterial translocation from the lumen, 156
progesterone, xi, 23, 27, 28, 29, 30, 31, 34, 36, 37, 39, 41, 43, 44, 45, 46, 47, 48, 49, 50, 59, 78, 84, 88
proliferative repair of the intestinal epithelium, 155
proposed molecular dynamics of HA intestinal uptake begins, 146
proteoglycans, 3, 12, 39, 54, 57, 100, 101, 103, 106, 109, 110, 112, 131, 135

R

radiation, xi, 36, 155
receptor(s), xi, 2, 3, 7, 8, 11, 12, 13, 15, 16, 17, 18, 20, 21, 22, 26, 27, 30, 32, 33, 34, 35, 36, 37, 39, 41, 43, 44, 46, 47, 49, 50, 54, 56, 59, 86, 87, 90, 99, 102, 105, 106, 108, 110, 113, 115, 117, 119, 124, 128, 129, 130, 131, 134, 135, 138, 139, 144, 145, 150,

151, 152, 158, 161, 163, 164, 165, 167, 169
reduction in systemic inflammation through upregulation of the anti-inflammatory cytokine, 155
reproduction, ix, 19, 39, 41, 47, 50, 55, 61, 62, 83, 85, 88, 91, 93, 97, 98, 101, 102, 103
RHAMM, 8, 13, 18, 26, 27, 102, 109, 113, 115, 124, 133, 134, 135, 138, 167

S

safety, ix, xii, 37, 65, 66, 67, 68, 69, 70, 71, 74, 75, 77, 81
skeletal development, 121, 122
synthase(s), 6, 24, 27, 43, 49, 58, 105, 107, 124, 144, 153
synthesis, 4, 9, 24, 25, 26, 29, 31, 32, 33, 42, 45, 48, 55, 57, 58, 77, 86, 90, 97, 105, 107, 110, 111, 112, 113, 123, 135, 137, 144, 145, 148, 151, 152, 155

synthesis and degradation, 4, 9

T

through amniotic fluid, 151
tight junction expression, 156
TLR, 27, 32, 33, 133
TLR2, 8, 20, 32, 110, 145, 167
TLR4, 8, 11, 18, 20, 32, 145, 146, 151, 153, 154, 165, 166, 169
TMEM2, 108, 110, 113, 119, 128, 137, 139, 140, 144
TSG6, 108, 110, 114, 115, 116, 117, 121, 150

V

versican, 62, 109, 111, 112, 120, 121, 122, 127, 129, 130, 132, 133, 134, 135, 136
villi (crypts) proliferate, 149